GUIDE
FOR THE CARE AND USE OF
LABORATORY ANIMALS

Eighth Edition

Committee for the Update of the Guide for the Care and Use of Laboratory Animals

Institute for Laboratory Animal Research

Division on Earth and Life Studies

NATIONAL RESEARCH COUNCIL
OF THE NATIONAL ACADEMIES

THE NATIONAL ACADEMIES PRESS
Washington, D.C.
www.nap.edu

THE NATIONAL ACADEMIES PRESS 500 Fifth Street, NW Washington, DC 20001

NOTICE: The project that is the subject of this report was approved by the Governing Board of the National Research Council, whose members are drawn from the councils of the National Academy of Sciences, the National Academy of Engineering, and the Institute of Medicine. The members of the Committee responsible for the report were chosen for their special competences and with regard for appropriate balance.

This study was supported by the Office of Extramural Research, Office of the Director, National Institutes of Health/Department of Health and Human Services under Contract Number N01-OD-4-2139 Task Order #188; the Office of Research Integrity, Department of Health and Human Services; the Animal and Plant Health Inspection Service, U.S. Department of Agriculture; Association for Assessment and Accreditation of Laboratory Animal Care International; American Association for Laboratory Animal Science; Abbott Fund; Pfizer; American College of Laboratory Animal Medicine; American Society of Laboratory Animal Practitioners; Association of Primate Veternarians.

International Standard Book Number-13: 978-0-309-15400-0 (Book)
International Standard Book Number-10: 0-309-15400-6 (Book)
International Standard Book Number-13: 978-0-309-15401-7 (PDF)
International Standard Book Number-10: 0-309-15401-4 (PDF)
Library of Congress Control Number: 2010940400

Additional copies of this report are available from the National Academies Press, 500 Fifth Street, NW, Lockbox 285, Washington, DC 20055; (800) 624-6242 or (202) 334-3313 (in the Washington metropolitan area); http://www.nap.edu.

Printed in the United States of America.

THE NATIONAL ACADEMIES

Advisers to the Nation on Science, Engineering, and Medicine

The **National Academy of Sciences** is a private, nonprofit, self-perpetuating society of distinguished scholars engaged in scientific and engineering research, dedicated to the furtherance of science and technology and to their use for the general welfare. Upon the authority of the charter granted to it by the Congress in 1863, the Academy has a mandate that requires it to advise the federal government on scientific and technical matters. Dr. Ralph J. Cicerone is president of the National Academy of Sciences.

The **National Academy of Engineering** was established in 1964, under the charter of the National Academy of Sciences, as a parallel organization of outstanding engineers. It is autonomous in its administration and in the selection of its members, sharing with the National Academy of Sciences the responsibility for advising the federal government. The National Academy of Engineering also sponsors engineering programs aimed at meeting national needs, encourages education and research, and recognizes the superior achievements of engineers. Dr. Charles Vest is president of the National Academy of Engineering.

The **Institute of Medicine** was established in 1970 by the National Academy of Sciences to secure the services of eminent members of appropriate professions in the examination of policy matters pertaining to the health of the public. The Institute acts under the responsibility given to the National Academy of Sciences by its congressional charter to be an adviser to the federal government and, upon its own initiative, to identify issues of medical care, research, and education. Dr. Harvey V. Fineberg is president of the Institute of Medicine.

The **National Research Council** was organized by the National Academy of Sciences in 1916 to associate the broad community of science and technology with the Academy's purposes of furthering knowledge and advising the federal government. Functioning in accordance with general policies determined by the Academy, the Council has become the principal operating agency of both the National Academy of Sciences and the National Academy of Engineering in providing services to the government, the public, and the scientific and engineering communities. The Council is administered jointly by both Academies and the Institute of Medicine. Dr. Ralph J. Cicerone and Dr. Charles M. Vest are chair and vice chair, respectively, of the National Research Council.

www.national-academies.org

COMMITTEE FOR THE UPDATE OF THE GUIDE FOR THE CARE AND USE OF LABORATORY ANIMALS

Members

Janet C. Garber (*Chair*), Garber Consulting
R. Wayne Barbee, Virginia Commonwealth University
Joseph T. Bielitzki, University of Central Florida
Leigh Ann Clayton, National Aquarium, Baltimore
John C. Donovan, BioResources, Inc.
Coenraad F. M. Hendriksen, Netherlands Vaccine Institute, Bilthoven, The Netherlands (until March 2009)
Dennis F. Kohn, Columbia University (retired)
Neil S. Lipman, Memorial Sloan-Kettering Cancer Center and Weill Cornell Medical College
Paul A. Locke, Johns Hopkins Bloomberg School of Public Health
John Melcher, U.S. Senate (retired)
Fred W. Quimby, Rockefeller University (retired)
Patricia V. Turner, University of Guelph, Canada
Geoffrey A. Wood, University of Guelph, Canada
Hanno Würbel, Justus Liebig University of Giessen, Germany

Staff

Lida Anestidou, Study Director
Frances Sharples, Acting Director
Kathleen Beil, Administrative Coordinator
Cameron H. Fletcher, Senior Editor
Ruth Crossgrove, Senior Editor
Radiah Rose, Manager of Editorial Projects
Rhonda Haycraft, Senior Project Assistant
Joanne Zurlo, Director (until April 2010)

INSTITUTE FOR LABORATORY ANIMAL RESEARCH COUNCIL

Staff

Frances Sharples, Acting Director
Lida Anestidou, Senior Program Officer
Kathleen Beil, Administrative Coordinator
Cameron H. Fletcher, Managing Editor, *ILAR Journal*
Rhonda Haycraft, Program Associate
Joanne Zurlo, Director (until April 2010)

INSTITUTE FOR LABORATORY ANIMAL RESEARCH PUBLICATIONS

Recognition and Alleviation of Pain in Laboratory Animals (2009)
Scientific and Humane Issues in the Use of Random Source Dogs and Cats for Research (2009)
Recognition and Alleviation of Distress in Laboratory Animals (2008)
Toxicity Testing in the 21st Century: A Vision and a Strategy (2007)
Overcoming Challenges to Develop Countermeasures Against Aerosolized Bioterrorism Agents: Appropriate Use of Animal Models (2006)
Guidelines for the Humane Transportation of Research Animals (2006)
Science, Medicine, and Animals: Teacher's Guide (2005)
Animal Care and Management at the National Zoo: Final Report (2005)
Science, Medicine, and Animals (2004)
The Development of Science-based Guidelines for Laboratory Animal Care: Proceedings of the November 2003 International Workshop (2004)
Animal Care and Management at the National Zoo: Interim Report (2004)
National Need and Priorities for Veterinarians in Biomedical Research (2004)
Guidelines for the Care and Use of Mammals in Neuroscience and Behavioral Research (2003)
International Perspectives: The Future of Nonhuman Primate Resources, Proceedings of the Workshop Held April 17-19, 2002 (2003)
Occupational Health and Safety in the Care and Use of Nonhuman Primates (2003)
Definition of Pain and Distress and Reporting Requirements for Laboratory Animals: Proceedings of the Workshop Held June 22, 2000 (2000)
Strategies That Influence Cost Containment in Animal Research Facilities (2000)
Microbial Status and Genetic Evaluation of Mice and Rats: Proceedings of the 1999 US/Japan Conference (2000)
Microbial and Phenotypic Definition of Rats and Mice: Proceedings of the 1998 US/Japan Conference (1999)
Monoclonal Antibody Production (1999)
The Psychological Well-Being of Nonhuman Primates (1998)
Biomedical Models and Resources: Current Needs and Future Opportunities (1998)
Approaches to Cost Recovery for Animal Research: Implications for Science, Animals, Research Competitiveness and Regulatory Compliance (1998)
Chimpanzees in Research: Strategies for Their Ethical Care, Management, and Use (1997)

Occupational Health and Safety in the Care and Use of Research Animals (1997)
Guide for the Care and Use of Laboratory Animals (1996)
Rodents (1996)
Nutrient Requirements of Laboratory Animals, Fourth Revised Edition (1995)
Laboratory Animal Management: Dogs (1994)
Recognition and Alleviation of Pain and Distress in Laboratory Animals (1992)
Education and Training in the Care and Use of Laboratory Animals: A Guide for Developing Institutional Programs (1991)
Companion Guide to Infectious Diseases of Mice and Rats (1991)
Infectious Diseases of Mice and Rats (1991)
Immunodeficient Rodents: A Guide to Their Immunobiology, Husbandry, and Use (1989)
Use of Laboratory Animals in Biomedical and Behavioral Research (1988)
Animals for Research: A Directory of Sources, Tenth Edition and Supplement (1979)
Amphibians: Guidelines for the Breeding, Care and Management of Laboratory Animals (1974)

Copies of these reports may be ordered from the National Academies Press
(800) 624-6242 or (202) 334-3313
www.nap.edu

Reviewers

This eighth edition of the *Guide for the Care and Use of Laboratory Animals* has been reviewed in draft form by individuals chosen for their diverse perspectives and expertise, in accordance with procedures approved by the Report Review Committee of the National Research Council. The purpose of this independent review is to provide candid and critical comments that will assist the Committee in making its published report as sound as possible, and to ensure that the report meets institutional standards for objectivity, evidence, and responsiveness to the study charge. The review comments and draft manuscript remain confidential to protect the integrity of the deliberation process. The Committee thanks the following individuals for their review of the draft report:

Michael B. Ballinger, Amgen
Philippe J.R. Baneux, PreLabs
Stephen W. Barthold, University of California-Davis
Linda C. Cork, Stanford University
Jann Hau, University of Copenhagen, Denmark
Michael J. Huerkamp, Emory University
Michael D. Kastello, sanofi-aventis
Arthur L. Lage, Harvard Medical School
Christian Lawrence, Children's Hospital Boston
Randall J. Nelson, University of Tennessee College of Medicine-Memphis
Steven M. Niemi, Massachusetts General Hospital
Melinda A. Novak, University of Massachusetts-Amherst

Gemma Perretta, National Research Council, Italy
Marky E. Pitts, IACUC Consultant
George E. Sanders, University of Washington
Allen W. Singer, Battelle Memorial Institute
William J. White, Charles River Laboratories

Although the reviewers listed above have provided many constructive comments and suggestions, they were not asked to endorse the conclusions or recommendations nor did they see the final draft of the report before its release. The review of this report was overseen by John Dowling, Harvard University, and John Vandenbergh, North Carolina State University. Appointed by the National Research Council, they were responsible for making certain that an independent examination of this report was carried out in accordance with institutional procedures and that all review comments were carefully considered. Responsibility for the final content of this report rests entirely with the authoring committee and the institution.

Preface

The purpose of the *Guide for the Care and Use of Laboratory Animals* (the *Guide*), as expressed in the charge to the Committee for the Update of the *Guide*, is to assist institutions in caring for and using animals in ways judged to be scientifically, technically, and humanely appropriate. The *Guide* is also intended to assist investigators in fulfilling their obligation to plan and conduct animal experiments in accord with the highest scientific, humane, and ethical principles. Recommendations in the *Guide* are based on published data, scientific principles, expert opinion, and experience with methods and practices that have proved to be consistent with both high-quality research and humane animal care and use. These recommendations should be used as a foundation for the development of a comprehensive animal care and use program, recognizing that the concept and application of performance standards, in accordance with goals, outcomes, and considerations defined in the *Guide*, is essential to this process.

The *Guide* is an internationally accepted primary reference on animal care and use, and its use is required in the United States by the Public Health Service Policy. It was first published in 1963, under the title *Guide for Laboratory Animal Facilities and Care*, and was revised in 1965, 1968, 1972, 1978, 1985, and 1996. More than 550,000 copies have been printed since its first publication.

In 2006 an ad hoc committee appointed by the Institute for Laboratory Animal Research recommended that the *Guide* be updated. The Committee for the Update of the *Guide for the Care and Use of Laboratory Animals* was appointed in 2008 by the National Research Council; its 13 members included research scientists, veterinarians, and nonscientists representing biomedical ethics and the public's interest in animal welfare. The Committee widely solicited written and oral comments on the update of the *Guide* from the scientific community and the general public; comments at open meetings (on September 26, 2008, in Washington, DC; October 16, 2008, in Irvine, California; and November 14, 2008, in Chicago) as well as written comments submitted to or requested by the Committee were considered. In addition, the Committee studied the materials submitted to NIH in response to its 2005 Request for Information (NOT-OD-06-011). All comments contributed substantially to this eighth edition of the *Guide*.

In approaching its task, the Committee carried forward the balance between ethical and science-based practice that has always been the basis of the *Guide*, and fulfilled its role to provide an updated resource that enables the research community to proceed responsibly and in a self-regulatory manner with animal experimentation. The *Guide* is predicated on the understanding that the exercise of professional judgment both upholds the central notion of performance standards and obviates the need for more stringent regulations.

Laboratory animal science is a rapidly evolving field and the Committee identified a number of areas in which current available scientific information is insufficient; additional objective information and assessment are needed to provide a scientific basis for recommendations in future editions of the *Guide*. Although pursuing these concepts was beyond this Committee's charge, the following two topics merit further study: (1) space and housing needs of laboratory species and (2) the need and best methods for providing enrichment, exercise, and human contact.

The need for continual updating of the *Guide* is implicit in its objective "to provide information that will enhance animal well-being, the quality of research, and the advancement of scientific knowledge that is relevant to both humans and animals" (Chapter 1). The irregular and increasing intervals between updates, reaching a 14-year gap between the seventh edition and this eighth edition, mean that important new research findings might wait more than a decade before being reflected in recommended practice. Addressing this concern was beyond the charge of this Committee; we noted, however, that regular and more frequent updates of the information in the *Guide* will promote laboratory animal welfare and support high-quality scientific data. A formal process for revising the information in the *Guide*, including the updating of practice standards, could meet this need.

In undertaking this update, the Committee acknowledged the contributions of William I. Gay and Bennett J. Cohen in the development of the original *Guide*. In 1959, Animal Care Panel (ACP) President Cohen appointed the Committee on Ethical Considerations in the Care of Laboratory Animals to evaluate animal care and use. That Committee was chaired by Dr. Gay, who soon recognized that the Committee could not evaluate animal care programs objectively without appropriate criteria on which to base its evaluations—that is, standards were needed. The ACP Executive Committee agreed, and the Professional Standards Committee was appointed. NIH later awarded the ACP a contract to "determine and establish a professional standard for laboratory animal care and facilities." Dr. Cohen chaired the ACP Animal Facilities Standards Committee, which prepared the first *Guide for Laboratory Animal Facilities and Care.*

This edition of the *Guide* was financially supported by the National Institutes of Health; the Office of Research Integrity, Department of Health and Human Services; the US Department of Agriculture (USDA); the Association for Assessment and Accreditation of Laboratory Animal Care International; the American Association for Laboratory Animal Science; Abbott Fund; Pfizer, Inc.; the American College of Laboratory Animal Medicine; the American Society of Laboratory Animal Practitioners; and the Association of Primate Veterinarians.

The Committee for the Update of the *Guide for the Care and Use of Laboratory Animals* expresses its appreciation to the Animal Welfare Information Center, National Agricultural Library, USDA, for its assistance in compiling bibliographies and references. This task would have been formidable without the help of the Center's staff. Appreciation is also extended to the reviewers of this volume, to Rhonda Haycraft for providing exemplary administrative and logistical assistance, and especially to Lida Anestidou, Study Director, who, through extraordinary patience, persistence, and scientific insight, managed the process from beginning to end.

Readers who detect errors of omission or commission are invited to send corrections and suggestions to the Institute for Laboratory Animal Research, National Research Council, 500 Fifth Street NW, Washington, DC 20001.

Janet C. Garber, *Chair*
Committee for the Update of the *Guide for the Care and Use of Laboratory Animals*

Overview

This eighth edition of the *Guide* is divided into five chapters and four appendices.

Chapter 1 presents the goals and intended audiences of the *Guide* as well as key concepts and terminology essential to its premise and use. Incorporating some of the material from the Introduction to the last edition, the chapter highlights a commitment to the concepts of the Three Rs—Replacement, Reduction, and Refinement—and provides an enhanced discussion of the ethics of animal use and investigator/institutional obligations.

Chapter 2 focuses on the overall institutional *animal care and use program* (Program), in addition to many of the topics previously covered in Chapter 1 of the seventh edition. It defines the evolved concept of Program and provides a framework for its intra-institutional integration, taking into account institutional policies and responsibilities, regulatory considerations, Program and personnel management (including training and occupational health and safety), and Program oversight. Discussions of the latter include institutional animal care and use committee (IACUC) functions, protocol and Program review, postapproval monitoring (a new section), and considerations such as humane endpoints and multiple survival surgical procedures. The Committee endorses the American College of Laboratory Animal Medicine's "Guidelines for Adequate Veterinary Care."

Chapter 3 focuses on the animals themselves and, unlike previous editions, addresses terrestrial and aquatic species in separate sections, reflecting the growing role of aquatic animals in biomedical research. The chapter provides recommendations for housing and environment, discusses the importance of social housing, and includes enhanced sections on environmental enrichment, animal well-being, and scientific validity.

Space recommendations were nominally expanded based on the Committee's professional and expert opinion and on current housing methods. Cage sizes have historically been interpreted as minimum space needs by users of the *Guide*, and were labeled as such ("recommended minimum space") in this edition. The use of the word "minimum" does not further restrict users of the *Guide* because, although the space requirements are numbers (i.e., engineering standards), they are used in a performance standards framework. The Committee recommends minimum space for female rodents with litter and an increase of the cage height for rabbits to 16". Further, in light of many comments submitted to the Committee requesting more information on performance goals and how to achieve them, rodent breeding recommendations are accompanied by substantial guidance.

With respect to nonhuman primates (NHPs), the Committee endorses social housing as the default and has provided some species-specific guidance. An additional group has been added for monkeys, and chimpanzees are separated in a new category. These changes were motivated by the Committee's recognition (affirmed in comments solicited from NHP experts) that these animals need more floor and vertical space, at least in some groups, to exercise their natural habits.

Chapter 4 discusses veterinary care and the responsibilities of the attending veterinarian. It introduces the concept of *animal biosecurity* and upholds its central role in ensuring the health of laboratory animals. The chapter includes recommendations relative to animal procurement, transportation, and preventive medicine, and expands the sections on clinical care and management, surgery (with a new section on intraoperative monitoring), pain and distress, and euthanasia.

Chapter 5 discusses physical plant–related topics and includes updated and new material on vibration control; physical security and access control; hazardous agent containment; and special facilities for imaging and whole body irradiation, barrier housing, behavioral studies, and aquatic species housing. The chapter provides detailed discussion of centralized versus decentralized animal facilities and introduces the concept of variable-volume HVAC systems with a nod toward energy conservation and efficiency.

Appendix A is the updated bibliography; Appendix B presents the U.S. Government Principles for the Utilization and Care of Vertebrate Animals Used in Testing, Research, and Training; Appendix C presents the Statement of Task; and Appendix D provides the biographical sketches of the Committee members.

In accordance with the Statement of Task ("In addition to the published report, the updated *Guide* will be posted on the Internet in a pdf or equivalent format such that users will be able to search the entire document at one time"), the *Guide* is available in searchable pdf format on the National Academies Press website, www.nap.edu.

Contents

1

Key Concepts

This edition of the *Guide for the Care and Use of Laboratory Animals* (the *Guide*) strongly affirms the principle that all who care for, use, or produce animals for research, testing, or teaching must assume responsibility for their well-being. The *Guide* is created by scientists and veterinarians for scientists and veterinarians to uphold the scientific rigor and integrity of biomedical research with laboratory animals as expected by their colleagues and society at large.

The *Guide* plays an important role in decision making regarding the use of vertebrate laboratory animals because it establishes the minimum ethical, practice, and care standards for researchers and their institutions. The use of laboratory animals in research, teaching, testing, and production is also governed or affected by various federal and local laws, regulations, and standards; for example, in the United States the Animal Welfare Act (AWA 1990) and Regulations (PL 89-544; USDA 1985) and/or Public Health Service (PHS) Policy (PHS 2002) may apply. Compliance with these laws, regulations, policies, and standards (or subsequent revised versions) in the establishment and implementation of a program of animal care and use is discussed in Chapter 2.

Taken together, the practical effect of these laws, regulations, and policies is to establish a system of self-regulation and regulatory oversight that binds researchers and institutions using animals. Both researchers and institutions have affirmative duties of humane care and use that are supported by practical, ethical, and scientific principles. This system of self-regulation establishes a rigorous program of animal care and use and provides flexibility in fulfilling the responsibility to provide humane care. The specific scope and nature of this responsibility can vary based on the scientific discipline, nature of the animal use, and species involved, but because it affects animal care and use in every situation this responsibility requires that producers, teachers, researchers, and institutions carry out purposeful analyses of proposed uses of laboratory animals. The *Guide* is central to these analyses and to the development of a program in which humane care is incorporated into all aspects of laboratory animal care and use.

APPLICABILITY AND GOALS

In the *Guide, laboratory animals* (also referred to as *animals*) are generally defined as any vertebrate animal (i.e., traditional laboratory animals, agricultural animals, wildlife, and aquatic species) produced for or used in research, testing, or teaching. *Animal use* is defined as the proper care, use, and humane treatment of laboratory animals produced for or used in research, testing, or teaching.

> *Laboratory animals* or *animals:* Any vertebrate animal (e.g., traditional laboratory animals, agricultural animals, wildlife, and aquatic species) produced for or used in research, testing, or teaching.

When appropriate, considerations or specific emphases for agricultural animals and nontraditional species are presented. The *Guide* does not address in detail agricultural animals used in production, agricultural research or teaching, wildlife and aquatic species studied in natural settings, or invertebrate animals (e.g., cephalopods) used in research, but establishes general principles and ethical considerations that are also applicable to these species and situations. References provide the reader with additional resources, and supplemental information on breeding, care, management, and use of selected laboratory animal species is available in other publications prepared by the Institute for Laboratory Animal Research (ILAR) and other organizations (Appendix A).

> *Animal use:* The proper care, use, and humane treatment of laboratory animals produced for or used in research, testing, or teaching.

The goal of the *Guide* is to promote the humane care and use of laboratory animals by providing information that will enhance animal well-being, the quality of research, and the advancement of scientific knowledge that is relevant to both humans and animals. The Committee recognizes that the use of different species in research is expanding and that researchers and institutions will face new and unique challenges in determining how to apply the *Guide* in these situations. In making such determinations, it is important to keep in mind that the *Guide* is intended to provide information to assist researchers, institutional animal care and use committees (IACUCs), veterinarians, and other stakeholders in ensuring the implementation of effective and appropriate animal care and use programs that are based on humane care. Throughout the *Guide*, scientists and institutions are encouraged to give careful and deliberate thought to the decision to use animals, taking into consideration the contribution that such use will make to new knowledge, ethical concerns, and the availability of alternatives to animal use (NRC 1992). A practical strategy for decision making, the "Three Rs" (Replacement, Reduction, and Refinement) approach, is discussed in more detail below. Institutions should use the recommendations in the *Guide* as a foundation for the development of a comprehensive animal care and use program and a process for continually improving this program.

INTENDED AUDIENCES AND USES OF THE *GUIDE*

The *Guide* is intended for a wide and diverse audience, including

- the scientific community
- administrators
- IACUCs
- veterinarians
- educators and trainers
- producers of laboratory animals
- accreditation bodies
- regulators
- the public.

The *Guide* is meant to be read by the user in its entirety, as there are many concepts throughout that may be helpful. Individual sections will be particularly relevant to certain users, and it is expected that the reader will explore in more detail the references provided (including those in Appendix A) on topics of interest.

Members of the scientific community (investigators and other animal users) will find Chapters 1 and 2 (and portions of Chapter 4) of the *Guide* useful for their interactions with the IACUC, attending veterinarian, and administrators regarding animal care as well as the preparation of animal care and use protocols. Scientific review committees and journal editors may choose to refer to multiple sections of the *Guide* to determine whether scientists contributing proposals and manuscripts have met the appropriate standards in their planned use of animals. The *Guide* can assist IACUCs and administrators in protocol review, assessment, and oversight of an animal care and use program. Veterinarians should find Chapters 3 through 5 valuable for their oversight and support of animal care and use. Educators and trainers can use the *Guide* as a document to assess both the scope and adequacy of training programs supported by the institution. Accreditation bodies will find the *Guide* useful for evaluating many areas of animal care and use programs not subject to strict engineering standards (see definition below). Finally, members of the public should feel assured that adherence to the *Guide* will ensure humane care and use of laboratory animals.

Readers are reminded that the *Guide* is used by a diverse group of national and international institutions and organizations, many of which are covered by neither the Animal Welfare Act nor the PHS Policy. The *Guide* uses some terminology that is both defined by US statute and denotes a general concept (e.g., "attending veterinarian," "adequate veterinary care," and "institutional official"). Even if these terms are not consistent with those used by non-US institutions, the underlying principles can still be applied. In all instances where *Guide* recommendations are different from applicable legal or policy requirements, the higher standard should apply.

ETHICS AND ANIMAL USE

The decision to use animals in research requires critical thought, judgment, and analysis. Using animals in research is a privilege granted by society to the research community with the expectation that such use will provide either significant new knowledge or lead to improvement in human and/or animal well-being (McCarthy 1999; Perry 2007). It is a trust that mandates responsible and humane care and use of these animals. The *Guide* endorses the responsibilities of investigators as stated in the *U.S. Government Principles for Utilization and Care of Vertebrate Animals Used in Testing, Research, and Training* (IRAC 1985; see Appendix B). These principles direct the research community to accept responsibility for the care and use of animals during all phases of the research effort. Other government agencies and professional organizations have published similar principles (NASA 2008; NCB 2005; NIH 2006, 2007; for additional references see Appendix A). Ethical considerations discussed here and in other sections of the *Guide* should serve as a starting point; readers are encouraged to go beyond these provisions. In certain situations, special considerations will arise during protocol review and planning; several of these situations are discussed in more detail in Chapter 2.

THE THREE Rs

The Three Rs represent a practical method for implementation of the principles described above. In 1959, W.M.S. Russell and R.L. Burch published a practical strategy of replacement, refinement, and reduction—referred to as the Three Rs—for researchers to apply when considering experimental design in laboratory animal research (Russell and Burch 1959). Over the years, the Three Rs have become an internationally accepted approach for researchers to apply when deciding to use animals in research and in designing humane animal research studies.

Replacement refers to methods that avoid using animals. The term includes absolute replacements (i.e., replacing animals with inanimate systems such as computer programs) as well as relative replacements (i.e., replacing animals such as vertebrates with animals that are lower on the phylogenetic scale).

Refinement refers to modifications of husbandry or experimental procedures to enhance animal well-being and minimize or eliminate pain and distress. While institutions and investigators should take all reasonable measures to eliminate pain and distress through refinement, IACUCs should understand that with some types of studies there may be either unforeseen or intended experimental outcomes that produce pain. These outcomes may or may not be eliminated based on the goals of the study.

Reduction involves strategies for obtaining comparable levels of information from the use of fewer animals or for maximizing the information obtained from a given number of animals (without increasing pain or distress) so that in the long run fewer animals are needed to acquire the same scientific information. This approach relies on an analysis of experimental design, applications of newer technologies, the use of appropriate statistical methods, and control of environmentally related variability in animal housing and study areas (see Appendix A).

Refinement and reduction goals should be balanced on a case-by-case basis. Principal investigators are strongly discouraged from advocating animal reuse as a reduction strategy, and reduction should not be a rationale for reusing an animal or animals that have already undergone experimental procedures especially if the well-being of the animals would be compromised. Studies that may result in severe or chronic pain or significant alterations in the animals' ability to maintain normal physiology, or adequately respond to stressors, should include descriptions of appropriate humane endpoints or provide science-based justification for not using a particular, commonly accepted humane endpoint. Veterinary consultation must occur when pain or distress is beyond the level anticipated in the protocol description or when interventional control is not possible.

KEY TERMS USED IN THE *GUIDE*

The Committee for the Update of the *Guide* believes that the terms set out below are important for a full understanding of the *Guide*. Accordingly, we have defined these terms and concepts to provide users of the *Guide* with additional assistance in implementing their responsibilities.

Humane Care

Humane care means those actions taken to ensure that laboratory animals are treated according to high ethical and scientific standards. Implementation of a humane care program, and creation of a laboratory environment in which humane care and respect for animals are valued and encouraged, underlies the core requirements of the *Guide* and the system of self-regulation it supports (Klein and Bayne 2007).

Animal Care and Use Program

The *animal care and use program* (the Program) means the policies, procedures, standards, organizational structure, staffing, facilities, and practices put into place by an institution to achieve the humane care and use of animals in the laboratory and throughout the institution. It includes the establishment and support of an IACUC or equivalent ethical oversight committee and the maintenance of an environment in which the IACUC can function successfully to carry out its responsibilities under the *Guide* and applicable laws and policies. Chapter 2 provides a more expansive discussion of the importance of the *Guide* and its application to animal care and use programs.

Engineering, Performance, and Practice Standards

Engineering standard means a standard or guideline that specifies in detail a method, technology, or technique for achieving a desired outcome; it does not provide for modification in the event that acceptable alternative methods are available or unusual circumstances arise. Engineering standards are prescriptive and provide limited flexibility for implementation. However, an engineering standard can be useful to establish a baseline and is relatively easy to use in evaluating compliance.

Performance standard means a standard or guideline that, while describing a desired outcome, provides flexibility in achieving this outcome by granting discretion to those responsible for managing the animal care and use program, the researcher, and the IACUC. The performance approach requires professional input, sound judgment, and a team approach to achieve specific goals. It is essential that the desired outcomes and/or goals be clearly defined and appropriate performance measures regularly monitored in order to verify the success of the process. Performance standards can be advantageous because they accommodate the consideration of many variables (such as the species and previous history of the animals, facilities, staff expertise, and research goals) so that implementation can be best tailored to meet the recommendations in the *Guide*.

Ideally, engineering and performance standards are balanced, setting a target for optimal practices, management, and operations while encouraging flexibility and judgment, if appropriate, based on individual situations (Gonder et al. 2001).

Scientists, veterinarians, technicians, and others have extensive experience and information covering many of the topics discussed in the *Guide*. For topics on which information is insufficient or incomplete, sustained research into improved methods of laboratory animal management, care, and use is needed for the continued evaluation and improvement of performance and engineering standards.

Practice standard means the application of professional judgment by qualified, experienced individuals to a task or process over time, an approach that has been demonstrated to benefit or enhance animal care and use. Professional judgment comes from information in the peer-reviewed scientific literature and textbooks and, as in many other disciplines, from time-proven experiences in the field (for additional information see Chapter 2). In the absence of published scientific literature or other definitive sources, where experience has demonstrated that a particular practice improves animal care and use, practice standards have been used in determining appropriate recommendations in the *Guide*. In most situations, the *Guide* is intended to provide flexibility so that institutions can modify practices and procedures with changing conditions and new information.

POLICIES, PRINCIPLES, AND PROCEDURES

Policies commonly derive from a public agency or private entity. They are generally practical statements of collective wisdom, convention, or management direction that are internal to the entity. However, policies may assume broader force when they become the means by which an implementing agency interprets existing statutes (e.g., PHS Policy). *Principles* are broader in their scope and intended application, and are accepted generalizations about a topic that are frequently endorsed by many and diverse organizations (e.g., the U.S. Government Principles). *Procedures* (often called "operating procedures" or "standard operating procedures") are typically detailed, step-by-step processes meant to ensure the consistent application of institutional practices. Establishing standard operating procedures can assist an institution in complying with regulations, policies, and principles as well as with day-to-day operations and management.

MUST, SHOULD, AND MAY

Must indicates actions that the Committee for the Update of the *Guide* considers imperative and mandatory duty or requirement for providing humane animal care and use. *Should* indicates a strong recommendation for achieving a goal; however, the Committee recognizes that individual circumstances might justify an alternative strategy. *May* indicates a suggestion to be considered.

The *Guide* is written in general terms so that its recommendations can be applied in diverse institutions and settings that produce or use animals for research, teaching, and testing. This approach requires that users, IACUCs, veterinarians, and producers apply professional judgment in making specific decisions regarding animal care and use. Because the *Guide* is written in general terms, IACUCs have a key role in interpretation, implementation, oversight, and evaluation of institutional animal care and use programs.

REFERENCES

AWA [Animal Welfare Act]. 1990. Animal Welfare Act. PL (Public Law) 89-544. Available at www.nal.usda.gov/awic/legislat/awa.htm; accessed January 14, 2010.

Gonder JC, Smeby RR, Wolfle TL. 2001. Performance Standards and Animal Welfare: Definition, Application and Assessment, Parts I and II. Greenbelt MD: Scientists Center for Animal Welfare.

IRAC [Interagency Research Animal Committee]. 1985. U.S. Government Principles for Utilization and Care of Vertebrate Animals Used in Testing, Research, and Training. Federal Register, May 20, 1985. Washington: Office of Science and Technology Policy. Available at http://oacu.od.nih.gov/regs/USGovtPrncpl.htm; accessed May 10, 2010.

Klein HJ, Bayne KA. 2007. Establishing a culture of care, conscience, and responsibility: Addressing the improvement of scientific discovery and animal welfare through science-based performance standards. ILAR J 48:3-11.

McCarthy CR. 1999. Bioethics of laboratory animal research. ILAR J 40:1-37.

NASA [National Aeronautics and Space Administration]. 2008. NASA Principles for the Ethical Care and Use of Animals. NPR 8910.1B-Appendix A. May 28. Available at http://nodis3.gsfc.nasa.gov/displayDir.cfm?t=NPDandc=8910ands=1B; accessed May 10, 2010.

NCB [Nuffield Council on Bioethics]. 2005. The Ethics of Research Using Animals. London: NCB.

NIH [National Institutes of Health]. 2007. Memorandum of Understanding Between the Office of Laboratory Animal Welfare, National Institutes of Health, US Department of Health and Human Services and the Office of Research Oversight and the Office of Research and Development, Veterans Health Administration, US Department of Veterans Affairs Concerning Laboratory Animal Welfare. November 2007. Bethesda: Office of Extramural Research, NIH. Available at http://grants.nih.gov/grants/olaw/references/mou_olaw_va_2007_11.htm.

NIH. 2006. Memorandum of Understanding Among the Animal and Plant Health Inspection Service USDA and the Food and Drug Administration, US Department of Health and Human Services, and the National Institutes of Health Concerning Laboratory Animal Welfare. March 1, 2006. Bethesda: Office of Extramural Research, NIH. Available at http://grants.nih.gov/grants/olaw/references/finalmou.htm.

NRC [National Research Council]. 1992. Report on Responsible Science. Washington: National Academy Press.

Perry P. 2007. The ethics of animal research: A UK perspective. ILAR J 48:42-46.

PHS [Public Health Service]. 2002. Public Health Service Policy on Humane Care and Use of Laboratory Animals. Publication of the Department of Health and Human Services, National Institutes of Health, Office of Laboratory Animal Welfare. Available at http://grants.nih.gov/grants/olaw/references/phspol.htm; accessed June 9, 2010.

Russell WMS, Burch RL. 1959. The Principles of Humane Experimental Technique. London: Methuen and Co. [Reissued: 1992, Universities Federation for Animal Welfare, Herts, UK].

USDA [US Department of Agriculture]. 1985. 9 CFR 1A. (Title 9, Chapter 1, Subchapter A): Animal Welfare. Available at http://ecfr.gpoaccess.gov/cgi/t/text/text-idx?sid=8314313bd7adf2c9f1964e2d82a88d92andc=ecfrandtpl=/ecfrbrowse/Title09/9cfrv1_02.tpl; accessed January 14, 2010.

2

Animal Care and Use Program

The proper care and use of laboratory animals in research, testing, teaching, and production (*animal use*) require scientific and professional judgment based on the animals' needs and their intended use. An animal care and use program (hereafter referred to as the Program) comprises all activities conducted by and at an institution that have a direct impact on the well-being of animals, including animal and veterinary care, policies and procedures, personnel and program management and oversight, occupational health and safety, institutional animal care and use committee (IACUC) functions, and animal facility design and management.

> *Program:* The activities conducted by and at an institution that have a direct impact on the well-being of animals, including animal and veterinary care, policies and procedures, personnel and program management and oversight, occupational health and safety, IACUC functions, and animal facility design and management.

This chapter defines the overall Program and key oversight responsibilities and provides guidelines to aid in developing an effective Program. Chapters 3, 4, and 5 cover the details of Program components: environment, housing, and management; veterinary care; and physical plant, respectively. Each institution should establish and provide sufficient resources for a Program that is managed in accord with the *Guide* and in compliance with applicable regulations, policies, and guidelines.

REGULATIONS, POLICIES, AND PRINCIPLES

The use of laboratory animals is governed by an interrelated, dynamic system of regulations, policies, guidelines, and procedures. The *Guide* takes into consideration regulatory requirements relevant to many US-based activities, including the Animal Welfare Regulations (USDA 1985; US Code, 42 USC § 289d) and the Public Health Service Policy on Humane Care and Use of Laboratory Animals (PHS 2002). The use of the *Guide* by non-US entities also presumes adherence to all regulations relevant to the humane care and use of laboratory animals applicable in those locations. The *Guide* also takes into account the U.S. Government Principles for Utilization and Care of Vertebrate Animals Used in Testing, Research, and Training (IRAC 1985; see Appendix B) and endorses the following principles:

- consideration of alternatives (in vitro systems, computer simulations, and/or mathematical models) to reduce or replace the use of animals
- design and performance of procedures on the basis of relevance to human or animal health, advancement of knowledge, or the good of society
- use of appropriate species, quality, and number of animals
- avoidance or minimization of discomfort, distress, and pain
- use of appropriate sedation, analgesia, and anesthesia
- establishment of humane endpoints
- provision of adequate veterinary care
- provision of appropriate animal transportation and husbandry directed and performed by qualified persons
- conduct of experimentation on living animals exclusively by and/or under the close supervision of qualified and experienced personnel.

Interpretation and application of these principles and the *Guide* require knowledge, expertise, experience, and professional judgment. Programs should be operated in accord with the *Guide* and relevant regulations, policies, and principles. Also, institutions are encouraged to establish and periodically review written procedures to ensure consistent application of *Guide* standards. Supplemental information on various aspects of animal care and use is available in other publications prepared by the Institute for Laboratory Animal Research (ILAR) and other organizations (Appendix A). References in the *Guide* provide the reader with additional information that supports statements made in the *Guide*. In the absence of published literature, some information in the *Guide* is derived from currently accepted practice standards in laboratory animal science (see Chapter 1). The body of literature related to animal science and use of animals is constantly evolving, requiring Programs to remain current with the information and best practices.

PROGRAM MANAGEMENT

An effective Program requires clearly defined roles that align responsibility with regulatory and management authority. US federal law creates a statutory basis for the *institutional official* (IO), the *attending veterinarian* (AV), and the *institutional animal care and use committee* (IACUC). The *Guide* endorses these concepts as important operating principles for all US and non-US animal care and use programs. Effective leadership in and collaboration among these three components, which not only oversee but also support animal users, are necessary (Lowman 2008; Van Sluyters 2008). In addition, interactions with regulatory and funding agencies and accreditation organizations are an integral part of the Program.

As summarized here and discussed throughout the *Guide*, the primary oversight responsibilities in the Program rest with the IO, the AV, and the IACUC. Their roles fit in a defined organizational structure where the reporting relationships, authorities, and responsibilities of each are clearly defined and transparent. Together they establish policies and procedures, ensure regulatory compliance, monitor Program performance, and support high-quality science and humane animal use. A Program that includes these elements and establishes a balance among them has the best chance of efficiently using resources while attaining the highest standards of animal well-being and scientific quality (Bayne and Garnett 2008; Van Sluyters 2008).

Program Management Responsibility

The Institutional Official

The *institutional official* (IO) bears ultimate responsibility for the Program, although overall Program direction should be a shared responsibility among the IO, AV, and IACUC. The IO has the authority to allocate the resources needed to ensure the Program's overall effectiveness. Program needs should be clearly and regularly communicated to the IO by the AV, the IACUC, and others associated with the Program (e.g., facilities management staff, occupational health and safety personnel, scientists). As a

> *Institutional official:* The individual who, as a representative of senior administration, bears ultimate responsibility for the Program and is responsible for resource planning and ensuring alignment of Program goals with the institution's mission.

representative of senior administration, the IO is responsible for resource planning and ensuring the alignment of Program goals of quality animal care and use with the institution's mission.

The Attending Veterinarian

The *attending veterinarian* (AV) is responsible for the health and well-being of all laboratory animals used at the institution. The institution must provide the AV with sufficient authority, including access to all animals, and resources to manage the program of veterinary care. The AV should oversee other aspects of animal care and use (e.g., husbandry, housing) to ensure that the Program complies with the *Guide*.

> *Attending veterinarian:* The veterinarian responsible for the health and well-being of all laboratory animals used at the institution.

Institutional mission, programmatic goals, including the nature of animal use at the institution, and Program size will determine whether full-time, part-time, or consultative veterinary services are needed. If a full-time veterinarian is not available on site, a consulting or part-time veterinarian should be available in visits at intervals appropriate to programmatic needs. In such instances, there must be an individual with assigned responsibility for daily animal care and use and facility management. While institutions with large animal care and use programs may employ multiple veterinarians, the management of veterinary medicine, animal care, and facility operations by a single administrative unit is often an efficient mechanism to administer all aspects of the Program.

The *Guide* endorses the American College of Laboratory Animal Medicine's (ACLAM) "Guidelines for Adequate Veterinary Care" (ACLAM 1996). These guidelines include veterinary access to all animals and their medical records, regular veterinary visits to facilities where animals are or may be housed or used, provisions for appropriate and competent clinical, preventive, and emergency veterinary care, and a system for legal animal procurement and transportation. Other responsibilities of the AV are outlined in the Program Oversight section below and in later chapters. For a Program to work effectively, there should be clear and regular communication between the AV and the IACUC.

The Institutional Animal Care and Use Committee

The *IACUC* (or institutional equivalent) is responsible for assessment and oversight of the institution's Program components and facilities. It should have sufficient authority and resources (e.g., staff, training, computers and related equipment) to fulfill this responsibility. Detailed information on the role and function of the IACUC is provided later in this chapter.

Collaborations

Interinstitutional collaboration has the potential to create ambiguities about responsibility for animal care and use. In cases of such collaboration involving animal use (beyond animal transport), the participating institutions should have a formal written understanding (e.g., a contract, memorandum of understanding, or agreement) that addresses the responsibility for offsite animal care and use, animal ownership, and IACUC review and oversight (AAALAC 2003). In addition, IACUCs from the participating institutions may choose to review protocols for the work being conducted.

Personnel Management

Training and Education

All personnel involved with the care and use of animals must be adequately educated, trained, and/or qualified in basic principles of laboratory animal science to help ensure high-quality science and animal well-being. The number and qualifications of personnel required to conduct and support a Program depend on several factors, including the type and size of the institution, the administrative structure for providing adequate animal care, the characteristics of the physical plant, the number and species of animals maintained, and the nature of the research, testing, teaching, and production activities. Institutions are responsible for providing appropriate resources to support personnel training (Anderson 2007), and the IACUC is responsible for providing oversight and for evaluating the effectiveness of the training program (Foshay and Tinkey 2007). All Program personnel training should be documented.

Veterinary and Other Professional Staff Veterinarians providing clinical and/or Program oversight and support must have the experience, training, and expertise necessary to appropriately evaluate the health and well-being of the species used in the context of the animal use at the institution. Veterinarians providing broad Program direction should be trained or have relevant experience in laboratory animal facility administration and management. Depending on the scope of the Program, professionals with expertise in other specific areas may be needed—in, for example, facility design and renovation, human resource management, pathology of laboratory animals, comparative genomics, facility and equipment maintenance, diagnostic laboratory operations, and behavioral management. Laboratory animal science and medicine are rapidly changing and evolving disciplines. The institution should provide opportunities and support for regular professional development and continuing education to ensure both that professional staff are knowledgeable about the latest practices and procedures and that laboratory animals receive high-quality care (Colby et al. 2007).

Animal Care Personnel Personnel caring for animals should be appropriately trained (see Appendix A, Education), and the institution should provide for formal and/or on-the-job training to facilitate effective implementation of the Program and the humane care and use of animals. Staff should receive training and/or have the experience to complete the tasks for which they are responsible. According to the Program scope, personnel with expertise in various disciplines (e.g., animal husbandry, administration, veterinary medical technology) may be required.

There are a number of options for training animal care personnel and technicians (Pritt and Duffee 2007). Many colleges have accredited programs in veterinary technology (AVMA 2010); most are 2-year programs that award Associate of Science degrees, some are 4-year programs that award Bachelor of Science degrees. Nondegree training, via certification programs for laboratory animal technicians and technologists, is available from the American Association for Laboratory Animal Science (AALAS), and there are various commercially available training materials appropriate for self-guided study (Appendix A).

Personnel caring for laboratory animals should also regularly engage in continuing education activities and should be encouraged to participate in local and national laboratory animal science meetings and in other relevant professional organizations. On-the-job training, supplemented with institution-sponsored discussion and training programs and reference materials applicable to their jobs and the species in their care, should be provided to each employee responsible for animal care (Kreger 1995).

Coordinators of institutional training programs can seek assistance from the Animal Welfare Information Center (AWIC), the Laboratory Animal Welfare and Training Exchange (LAWTE), AALAS, and ILAR (NRC 1991). The *Guide to the Care and Use of Experimental Animals* by the Canadian Council on Animal Care (CCAC 1993) and guidelines from other countries are valuable additions to the libraries of laboratory animal scientists (Appendix A).

The Research Team The institution should provide appropriate education and training to members of research teams—including principal investigators, study directors, research technicians, postdoctoral fellows, students, and visiting scientists—to ensure that they have the necessary knowledge and expertise for the specific animal procedures proposed and the species used (Conarello and Shepard 2007). Training should be tailored to the particular needs of research groups; however, all research groups should receive training in animal care and use legislation, IACUC function, ethics of animal use and the concepts of the Three Rs, methods for reporting concerns about animal use, occupational health and safety issues pertaining to animal use, animal handling, aseptic surgical technique, anesthesia and analgesia, euthanasia, and other subjects, as required by statute. Continuing education programs should be offered to reinforce training and provide updates that reflect changes in technology, legislation, and other relevant areas. Frequency of training opportunities should ensure that all animal users have adequate training before beginning animal work.

The IACUC It is the institution's responsibility to ensure that IACUC members are provided with training opportunities to understand their work and role. Such training should include formal orientation to introduce new members to the institution's Program; relevant legislation, regulations, guidelines, and policies; animal facilities and laboratories where animal use occurs; and the processes of animal protocol and program review (Greene et al. 2007). Ongoing opportunities to enhance their understanding of animal care and use in science should also be provided. For example, IACUC members may meet with animal care personnel and research teams; be provided access to relevant journals, materials, and web-based training; and be given opportunities to attend meetings or workshops.

Occupational Health and Safety of Personnel

Each institution must establish and maintain an occupational health and safety program (OHSP) as an essential part of the overall Program of animal care and use (CFR 1984a,b,c; DHHS 2009; PHS 2002). The OHSP must be consistent with federal, state, and local regulations and should focus on maintaining a safe and healthy workplace (Gonder 2002; Newcomer 2002; OSHA 1998a). The nature of the OHSP will depend on the facility, research activities, hazards, and animal species involved. The National Research Council's publication *Occupational Health and Safety in the Care and Use of Research Animals* (NRC 1997) contains guidelines and references for establishing and maintaining an effective, comprehensive OHSP (also see Appendix A). An effective OHSP requires coordination between the research program (as represented by the investigator), the animal care and use Program (as represented by the AV, IO, and IACUC), the environmental health and safety program, occupational health services, and administration (e.g., human resources, finance, and facility maintenance personnel). Establishment of a safety committee may facilitate communication and promote ongoing evaluation of health and safety in the workplace. In some cases there is a regulatory requirement for such a committee. Operational and day-to-day responsibility for safety in the workplace resides with the laboratory or facility supervisor (e.g., principal investigator, facility director, or a staff veterinarian) and depends on safe work practices by all employees.

Control and Prevention Strategies A comprehensive OHSP should include a hierarchy of control and prevention strategies that begins with the identification of hazards and the assessment of risk associated with those hazards. Managing risk involves the following steps: first, the appropriate design and operation of facilities and use of appropriate safety equipment (engineering controls); second, the development of processes and standard operating procedures (SOPs; administrative controls); and finally, the provision of appropriate personal protective equipment (PPE) for employees. Special safety equipment should be used in combination with appropriate management and safety practices (NIH 2002; OSHA 1998a,b). Managing risk using these strategies requires that personnel be trained, maintain good personal hygiene, be knowledgeable about the hazards in their work environment, understand the proper selection and use of equipment, follow established procedures, and use the PPE provided.

Hazard Identification and Risk Assessment The institutional OHSP should identify potential hazards in the work environment and conduct a critical assessment of the associated risks. An effective OHSP ensures that the risks associated with the experimental use of animals are identified and reduced to minimal and acceptable levels. Hazard identification and risk assessment are ongoing processes that involve individuals qualified to assess dangers associated with the Program and implement commensurate safeguards. Health and safety specialists with knowledge in relevant disciplines should be involved in risk assessment and the development of procedures to manage such risks.

Potential hazards include experimental hazards such as biologic agents (e.g., infectious agents or toxins), chemical agents (e.g., carcinogens and mutagens), radiation (e.g., radionuclides, X-rays, lasers), and physical hazards (e.g., needles and syringes). The risks associated with unusual experimental conditions such as those encountered in field studies or wildlife research should also be addressed. Other potential hazards—such as animal bites, exposure to allergens, chemical cleaning agents, wet floors, cage washers and other equipment, lifting, ladder use, and zoonoses—that are inherent in or intrinsic to animal use should be identified and evaluated. Once potential hazards have been identified, a critical ongoing assessment of the associated risks should be conducted to determine appropriate strategies to minimize or manage the risks.

The extent and level of participation of personnel in the OHSP should be based on the hazards posed by the animals and materials used (the severity or seriousness of the hazard); the exposure intensity, duration, and frequency (prevalence of the hazard); to some extent, the susceptibility (e.g., immune status) of the personnel; and the history of occupational illness and injury in the particular workplace (Newcomer 2002; NRC 1997). Ongoing identification and evaluation of hazards call for periodic inspections and reporting of potential hazardous conditions or "near miss" incidents.

Facilities, Equipment, and Monitoring The facilities required to support the OHSP will vary depending on the scope and activities of the Program. Their design should preferentially use engineering controls and equipment to minimize exposure to anticipated hazards (also see Chapter 5). Because a high standard of personal cleanliness is essential, changing, washing, and showering facilities and supplies appropriate to the Program should be available.

Where biologic agents are used, the Centers for Disease Control and Prevention (CDC) and National Institutes of Health (NIH) publication *Biosafety in Microbiological and Biomedical Laboratories* (BMBL; DHHS 2009) and the USDA standards (USDA 2002) should be consulted for appropriate facility design and safety procedures. These design and safety features are based on the level of risk posed by the agents used. Special facilities and safety equipment may be needed to protect the animal care and investigative staff, other occupants of the facility, the public, animals, and the environment from exposure to hazardous biologic, chemical, and physical agents used in animal experimentation (DHHS 2009; Frasier and Talka 2005; NIH 2002). When necessary, these facilities should be separated from other animal housing and support areas, research and clinical laboratories, and patient care facilities. They should be appropriately identified and access to them limited to authorized personnel.

Facilities, equipment, and procedures should also be designed, selected, and developed to reduce the possibility of physical injury or health risk to personnel (NIOSH 1997a,b). Engineering controls and equipment that address the risk of ergonomic injury in activities such as the lifting of heavy equipment or animals should be considered (AVMA 2008). Those are also frequently used to limit or control personnel exposure to animal allergens (Harrison 2001; Huerkamp et al. 2009). The potential for repetitive motion injuries in animal facilities (e.g., maintenance of large rodent populations and other husbandry activities) should also be assessed.

The selection of appropriate animal housing systems requires professional knowledge and judgment and depends on the nature of the hazards in question, the types of animals used, the limitations or capabilities of the facilities, and the design of the experiments. Experimental animals should be housed so that possibly contaminated food and bedding, feces, and urine can be handled in a controlled manner. Appropriate facilities, equipment, and procedures should be used for bedding disposal. Safety equipment should be properly maintained and its function periodically validated. Appropriate methods should be used for assessing and monitoring exposure to potentially hazardous biologic, chemical, and physical agents where required (e.g., ionizing radiation) or where the possibility of exceeding permissible exposure limits exists (CFR 1984b).

Personnel Training As a general rule, safety depends on trained personnel who rigorously follow safe practices. Personnel at risk should be provided with clearly defined procedures and, in specific situations, personal protective equipment to safely conduct their duties, understand the hazards involved, and be proficient in implementing the required safeguards. They should be trained regarding zoonoses, chemical, biologic, and physical hazards (e.g., radiation and allergies), unusual conditions or agents that might be part of experimental procedures (e.g., the use of human tissue in immunocompromised animals), handling of waste materials, personal hygiene, the appropriate use of PPE, and other considerations (e.g., precautions to be taken during pregnancy, illness, or immunosuppression) as appropriate to the risk imposed by their workplace.

Personal Hygiene The use of good personal hygiene will often reduce the possibility of occupational injury and cross contamination. Appropriate policies should be established and enforced, and the institution should supply suitable attire and PPE (e.g., gloves, masks, face shields, head covers, coats, coveralls, shoes or shoe covers) for use in the animal facility and laboratories in which animals are used. Soiled attire should be disposed of, laundered, or decontaminated by the institution as appropriate, and may require that special provisions be implemented if outside vendors are used. Personnel should wash and/or disinfect their hands and change clothing as often as necessary to maintain good personal hygiene. Outer garments worn in the animal rooms should not be worn outside the animal facility unless covered (NRC 1997). Personnel should not be permitted to eat, drink, use tobacco products, apply cosmetics, or handle or apply contact lenses in rooms and laboratories where animals are housed or used (DHHS 2009; NRC 1997; OSHA 1998a).

Animal Experimentation Involving Hazards When selecting specific safeguards for animal experimentation with hazardous agents, careful attention should be given to procedures for animal care and housing, storage and distribution of the agents, dose preparation and administration, body fluid and tissue handling, waste and carcass disposal, items that might be used temporarily and removed from the site (e.g., written records, experimental devices, sample vials), and personal protection.

Institutions should have written policies and procedures governing experimentation with hazardous biologic, chemical, and physical agents. An oversight process (such as the use of a safety committee) should be developed to involve persons who are knowledgeable in the evaluation and safe use of hazardous materials or procedures and should include review of the procedures and facilities to be used for specific safety concerns. Formal safety programs should be established to assess hazards, determine the safeguards needed for their control, and ensure that staff have the necessary training and skills and that facilities are adequate for the safe conduct of the research. Technical support should be provided to monitor and ensure compliance with institutional safety policies. A collaborative approach involving the investigator and research team, attending veterinarian, animal care technician, and occupational health and safety professionals may enhance compliance.

The BMBL (DHHS 2009) and NRC (1997) recommend practices and procedures, safety equipment, and facility requirements for working with hazardous biologic agents and materials. Facilities that handle agents of unknown risk should consult with appropriate CDC personnel about hazard control and medical surveillance. The use of highly pathogenic "select agents and toxins" in research requires that institutions develop a program and procedures for procuring, maintaining, and disposing of these agents (CFR 1998, 2002a,b; NRC 2004; PL 107-56; PL 107-188; Richmond et al. 2003). The use of immunodeficient or genetically modified animals (GMAs) susceptible to or shedding human pathogens, the use of human tissues and cell lines, or any infectious disease model can lead to an increased risk to the health and safety of personnel working with the animals (Lassnig et al. 2005; NIH 2002).

Hazardous agents should be contained in the study environment, for example through the use of airflow control during the handling and administering of hazardous agents, necropsies on contaminated animals (CDC and NIH 2000), and work with chemical hazards (Thomann 2003). Waste anesthetic gases should be scavenged to limit exposure.

Personal Protection While engineering and administrative controls are the first considerations for the protection of personnel, PPE appropriate for the work environment, including clean institution-issued protective clothing, should be provided as often as necessary. Protective clothing and equipment should not be worn beyond the boundary of the hazardous agent work area or the animal facility (DHHS 2009). If appropriate, personnel should shower when they leave the animal care, procedure, or dose preparation areas. Personnel with potential exposure to hazardous agents or certain species should be provided with PPE appropriate to the situation (CFR 1984c); for example, personnel exposed to nonhuman primates should have PPE such as gloves, arm protectors, suitable face masks, face shields, and goggles (NRC 2003a). Hearing protection should be available in high-noise areas (OSHA 1998c). Personnel working in areas where they might be exposed to contaminated airborne particulate material or vapors should have suitable respiratory protection (Fechter 1995; McCullough 2000; OSHA 1998d), with respirator fit testing and training in the proper use and maintenance of the respirator (OSHA 1998d; Sargent and Gallo 2003).

Medical Evaluation and Preventive Medicine for Personnel Development and implementation of a program of medical evaluation and preventive medicine should involve input from trained health professionals, such as occupational health physicians and nurses. Confidentiality and other medical and legal factors must be considered in the context of appropriate federal, state, and local regulations (e.g., PL 104-191).

A preemployment health evaluation and/or a health history evaluation before work assignment is advisable to assess potential risks for individual employees. Periodic medical evaluations are advisable for personnel in specific risk categories. For example, personnel required to use respiratory protection may also require medical evaluation to ensure that they are physically and psychologically able to use the respirator properly (Sargent and Gallo 2003). An appropriate immunization schedule should be adopted. It is important to immunize animal care personnel against tetanus (NRC 1997), and preexposure immunization should be offered to people at risk of infection or exposure to specific agents such as rabies virus (e.g., if working with species at risk for infection) or hepatitis B virus (e.g., if working with human blood or human tissues, cell lines, or stocks). Vaccination is recommended if research is to be conducted on infectious diseases for which effective vaccines are available. More specific recommendations are available in the BMBL (DHHS 2009). Preemployment or preexposure serum collection is advisable only in specific circumstances as determined by an occupational health and safety professional (NRC 1997). In such cases, identification, traceability, retention, and storage conditions of samples should be considered, and the purpose for which the serum samples will be used must be consistent with applicable federal and state laws.

Laboratory animal allergy has become a significant issue for individuals in contact with laboratory animals (Bush and Stave 2003; Gordon 2001; Wolfle and Bush 2001; Wood 2001). The medical surveillance program should promote the early diagnosis of allergies (Bush 2001; Bush and Stave 2003; Seward 2001) and include evaluation of an individual's medical history for preexisting allergies. Personnel training should include information about laboratory animal allergies, preventive control measures, early recognition and reporting of allergy symptoms, and proper techniques for working with animals (Gordon et at. 1997; Schweitzer et al. 2003; Thulin et al. 2002). PPE should be used to supplement, not replace, engineering or process controls (Harrison 2001; Reeb-Whitaker et al. 1999). If PPE for respiratory protection is necessary, appropriate fit testing and training should be provided.

Zoonosis surveillance should be a part of an OHSP (DHHS 2009; NRC 1997). Personnel should be instructed to notify their supervisors of potential or known exposures and of suspected health hazards and illnesses. Non-human primate diseases that are transmissible to humans can be serious hazards (NRC 2003a). Animal technicians, veterinarians, investigators, students, research technicians, maintenance workers, and others who have contact with nonhuman primates or their tissues and body fluids or who have duties in nonhuman primate housing areas should be routinely screened for tuberculosis. Because of the potential for exposure to *Macacine herpesvirus* 1 (formerly *Cercopithecine herpesvirus* 1 or Herpes B virus), personnel who work with or handle biologic samples (blood and tissues) from macaques should have access to and be instructed in the use of bite and scratch emergency care stations (Cohen et al. 2002). Injuries associated with macaques, their tissues or body fluids, or caging and equipment with which the animals have had direct contact, should be carefully evaluated and appropriate postexposure treatment and follow-up implemented (ibid.; NRC 2003a).

Clear procedures should be established for reporting all accidents, bites, scratches, and allergic reactions (NRC 1997), and medical care for such incidents should be readily available (Cohen et al. 2002; DHHS 2009).

Personnel Security

While contingency plans normally address natural disasters, they should also take into account the threats that criminal activities such as personnel harassment and assault, facility trespassing, arson, and vandalism pose to laboratory animals, research personnel, equipment and facilities, and biomedical research at the institution. Preventive measures should be considered, including preemployment screening and physical and information technology security (Miller 2007).

Investigating and Reporting Animal Welfare Concerns

Safeguarding animal welfare is the responsibility of every individual associated with the Program. The institution must develop methods for reporting and investigating animal welfare concerns, and employees should be aware of the importance of and mechanisms for reporting animal welfare concerns. In the United States, responsibility for review and investigation of these concerns rests with the IO and the IACUC. Response to such reports should include communication of findings to the concerned employee(s), unless such concerns are reported anonymously; corrective actions if deemed necessary; and a report to the IO of the issue, findings, and actions taken. Reported concerns and any corrective actions taken should be documented.

Mechanisms for reporting concerns should be posted in prominent locations in the facility and on applicable institutional website(s) with instructions on how to report the concern and to whom. Multiple points of contact, including senior management, the IO, IACUC Chair, and AV, are recommended. The process should include a mechanism for anonymity, compliance with applicable whistleblower policies, nondiscrimination against the concerned/reporting party, and protection from reprisals.

Training and regular communication with employees (including personnel such as custodial, maintenance, and administrative staff, who are farther removed from the animal use) about the institution's animal use activities may reduce potential concerns.

PROGRAM OVERSIGHT

The Role of the IACUC

IACUC Constitution and Function

The responsibility of the IACUC is to oversee and routinely evaluate the Program. It is the institution's responsibility to provide suitable orientation, background materials, access to appropriate resources, and, if necessary, specific training to assist IACUC members in understanding their roles and responsibilities and evaluating issues brought before the committee.

Committee membership includes the following:

- a Doctor of Veterinary Medicine either certified (e.g., by ACLAM, ECLAM, JCLAM, KCLAM) or with training and experience in laboratory animal science and medicine or in the use of the species at the institution
- at least one practicing scientist experienced in research involving animals
- at least one member from a nonscientific background, drawn from inside or outside the institution
- at least one public member to represent general community interests in the proper care and use of animals.

Public members should not be laboratory animal users, affiliated in any way with the institution, or members of the immediate family of a person who is affiliated with the institution. The public member may receive compensation for participation and ancillary expenses (e.g., meals, parking, travel), but the amount should be sufficiently modest that it does not become a substantial source of income and thus risk compromising the member's association with the community and public at large.

For large institutions with many administrative units or departments, no more than three voting members should be associated with a single administrative unit (USDA 1985). The size of the institution and the nature and extent of the Program will determine the number of members of the committee and their terms of appointment. Institutions with broad research programs may need to choose scientists from a number of disciplines and experience to properly evaluate animal use protocols.

The committee is responsible for oversight and evaluation of the entire Program and its components as described in other sections of the *Guide*. Its oversight functions include review and approval of proposed animal use (protocol review) and of proposed significant changes to animal use; regular inspection of facilities and animal use areas; regular review of the Program; ongoing assessment of animal care and use; and establishment of a mechanism for receipt and review of concerns involving the care and use of animals at the institution. The committee must meet as often as necessary to fulfill its responsibilities, and records of committee meetings and results of deliberations should be maintained. Program review and facilities inspections should occur at least annually or more often as required (e.g., by the Animal Welfare Act and PHS Policy). After review and inspection, a written report (including any minority views) should be provided to the IO about the status of the Program.

Protocol Review

The animal use protocol is a detailed description of the proposed use of laboratory animals. The following topics should be considered in the preparation of the protocol by the researcher and its review by the IACUC:

- rationale and purpose of the proposed use of animals
- a clear and concise sequential description of the procedures involving the use of animals that is easily understood by all members of the committee
- availability or appropriateness of the use of less invasive procedures, other species, isolated organ preparation, cell or tissue culture, or computer simulation (see Appendix A, Alternatives)
- justification of the species and number of animals proposed; whenever possible, the number of animals and experimental group sizes should be statistically justified (e.g., provision of a power analysis; see Appendix A, Experimental Design and Statistics)

- unnecessary duplication of experiments
- nonstandard housing and husbandry requirements
- impact of the proposed procedures on the animals' well-being
- appropriate sedation, analgesia, and anesthesia (indices of pain or invasiveness might aid in the preparation and review of protocols; see Appendix A, Anesthesia, Pain, and Surgery)
- conduct of surgical procedures, including multiple operative procedures
- postprocedural care and observation (e.g., inclusion of post-treatment or postsurgical animal assessment forms)
- description and rationale for anticipated or selected endpoints
- criteria and process for timely intervention, removal of animals from a study, or euthanasia if painful or stressful outcomes are anticipated
- method of euthanasia or disposition of animals, including planning for care of long-lived species after study completion
- adequacy of training and experience of personnel in the procedures used, and roles and responsibilities of the personnel involved
- use of hazardous materials and provision of a safe working environment.

While the responsibility for scientific merit review normally lies outside the IACUC, the committee members should evaluate scientific elements of the protocol as they relate to the welfare and use of the animals. For example, hypothesis testing, sample size, group numbers, and adequacy of controls can relate directly to the prevention of unnecessary animal use or duplication of experiments. For some IACUC questions, input from outside experts may be advisable or necessary. In the absence of evidence of a formal scientific merit review, the IACUC may consider conducting or requesting such a review (Mann and Prentice 2004). IACUC members named in protocols or who have other conflicts must recuse themselves from decisions concerning these protocols.

At times, protocols include procedures that have not been previously encountered or that have the potential to cause pain or distress that cannot be reliably predicted or controlled. Relevant objective information about the procedures and the purpose of the study should be sought from the literature, veterinarians, investigators, and others knowledgeable about the effects on animals. If little is known about a specific procedure, limited pilot studies, designed to assess both the procedure's effects on the animals and the skills of the research team and conducted under IACUC oversight, are appropriate. General guidelines for protocol or method evaluation for some of these situations are provided below, but they may not apply in all instances.

Special Considerations for IACUC Review

Certain animal use protocols include procedures or approaches that require special consideration during the IACUC review process due to their potential for unrelieved pain or distress or other animal welfare concerns. The topics below are some of the most common requiring special IACUC consideration. For these and other areas the IACUC is obliged to weigh the objectives of the study against potential animal welfare concerns. By considering opportunities for refinement, the use of appropriate nonanimal alternatives, and the use of fewer animals, both the institution and the principal investigator (PI) can begin to address their shared obligations for humane animal care and use.

Experimental and Humane Endpoints The experimental endpoint of a study occurs when the scientific aims and objectives have been reached. The humane endpoint is the point at which pain or distress in an experimental animal is prevented, terminated, or relieved. The use of humane endpoints contributes to refinement by providing an alternative to experimental endpoints that result in unrelieved or severe animal pain and distress, including death. The humane endpoint should be relevant and reliable (Hendriksen and Steen 2000; Olfert and Godson 2000; Sass 2000; Stokes 2002). For many invasive experiments, the experimental and humane endpoints are closely linked (Wallace 2000) and should be carefully considered during IACUC protocol review. While all studies should employ endpoints that are humane, studies that commonly require special consideration include those that involve tumor models, infectious diseases, vaccine challenge, pain modeling, trauma, production of monoclonal antibodies, assessment of toxicologic effects, organ or system failure, and models of cardiovascular shock.

The PI, who has precise knowledge of both the objectives of the study and the proposed model, should identify, explain, and include in the animal use protocol a study endpoint that is both humane and scientifically sound. The identification of humane endpoints is often challenging, however, because multiple factors must be weighed, including the model, species (and sometimes strain or stock), animal health status, study objectives, institutional policy, regulatory requirements, and occasionally conflicting scientific literature. Determination of humane endpoints should involve the PI, the veterinarian, and the IACUC, and should be defined when possible before the start of the study (Olfert and Godson 2000; Stokes 2000).

Information that is critical to the IACUC's assessment of appropriate endpoint consideration in a protocol includes precise definition of the humane endpoint (including assessment criteria), the frequency of animal observation, training of personnel responsible for assessment and recognition of the humane endpoint, and the response required upon reaching the humane endpoint. An understanding of preemptive euthanasia (Toth 2000), behavioral or physiologic definitions of the moribund state (ibid.), and the use of study-specific animal assessment records (Morton 2000; Paster et al. 2009) can aid the PI and IACUC when considering or developing proposed endpoints. When novel studies are proposed or information for an alternative endpoint is lacking, the use of pilot studies is an effective method for identifying and defining humane endpoints and reaching consensus among the PI, IACUC, and veterinarian. A system for communication with the IACUC should be in place both during and after such studies. Numerous publications address specific proposals for the application and use of humane endpoints (e.g., CCAC 1998; ILAR 2000; OECD 1999; Toth 1997; UKCCCR 1997).

Unexpected Outcomes Fundamental to scientific inquiry is the investigation of novel experimental variables. Because of the potential for unexpected outcomes that may affect animal well-being when highly novel variables are introduced, more frequent monitoring of animals may be required. With their inherent potential for unanticipated phenotypes, GMAs are an example of models for which increased monitoring for unexpected outcomes could be implemented (Dennis 1999).

GMAs, particularly mice and fish, are important animal models, and new methods and combinations of genetic manipulation are constantly being developed (Gondo 2008). Regardless of whether genetic manipulation is targeted or random, the phenotype that initially results is often unpredictable and may lead to expected or unexpected outcomes that affect the animal's well-being or survival at any stage of life. For example, in some instances genetic modification has led to unforeseen immunodeficiency, requiring the GMA offspring to be held under specialized bioexclusion conditions (Mumphrey et al. 2007); and the promoter sequences used to direct expression of transgenes to specific tissues have varying degrees of specificity ("leakiness") that can lead to unanticipated phenotypes (Moorehead et al. 2003). These examples illustrate the diversity of unanticipated outcomes and emphasize the need for diligent monitoring and professional judgment to ensure the animals' well-being (Dennis 2000). The first offspring of a newly generated GMA line should be carefully observed from birth into early adulthood for signs of disease, pain, or distress. Investigators may find that the phenotype precludes breeding of particular genotypes or that unexpected infertility occurs, situations that could lead to increases in the numbers of animals used and revision of the animal use protocol. When the initial characterization of a GMA reveals a condition that negatively affects animal well-being, this should be reported to the IACUC, and more extensive analysis may be required to better define the phenotype (Brown et al. 2000; Crawley 1999; Dennis 2000). Such monitoring and reporting may help to determine whether proactive measures can circumvent or alleviate the impact of the genetic modification on the animal's well-being and to establish humane endpoints specific to the GMA line.

Physical Restraint Physical restraint is the use of manual or mechanical means to limit some or all of an animal's normal movement for the purpose of examination, collection of samples, drug administration, therapy, or experimental manipulation. Animals are restrained for brief periods, usually minutes, in many research applications.

Restraint devices should be suitable in size, design, and operation to minimize discomfort, pain, distress, and the potential for injury to the animal and the research staff. Dogs, nonhuman primates, and many other animals can be trained, through use of positive reinforcement techniques, to cooperate with research procedures or remain immobile for brief periods (Boissy et al. 2007; Laule et al. 2003; Meunier 2006; Prescott and Buchanan-Smith 2003; Reinhardt 1991, 1995; Sauceda and Schmidt 2000; Yeates and Main 2009).

Prolonged restraint, including chairing of nonhuman primates, should be avoided unless it is essential for achieving research objectives and is specifically approved by the IACUC (NRC 2003b). Systems that do not limit an animal's ability to make normal postural adjustments (e.g., subcutaneous implantation of osmotic minipumps in rodents, backpack-fitted infusion pumps in dogs and nonhuman primates, and free-stall housing for farm animals) should be used when compatible with protocol objectives. Animals that do not adapt to necessary restraint systems should be removed from the study. When restraint devices are used, they should be specifically designed to accomplish research goals that are impossible or impractical to accomplish by other means or to prevent injury to animals or personnel.

The following are important guidelines for restraint:

- Restraint devices should not be considered a normal method of housing, and must be justified in the animal use protocol.
- Restraint devices should not be used simply as a convenience in handling or managing animals.
- Alternatives to physical restraint should be considered.
- The period of restraint should be the minimum required to accomplish the research objectives.
- Animals to be placed in restraint devices should be given training (with positive reinforcement) to adapt to the equipment and personnel.
- Animals that fail to adapt should be removed from the study.
- Provision should be made for observation of the animal at appropriate intervals, as determined by the IACUC.

- Veterinary care must be provided if lesions or illnesses associated with restraint are observed. The presence of lesions, illness, or severe behavioral change often necessitates the temporary or permanent removal of the animal from restraint.
- The purpose of the restraint and its duration should be clearly explained to personnel involved with the study.

Multiple Survival Surgical Procedures Surgical procedures in the laboratory setting may be categorized as major or minor (USDA 1985). Whether a procedure is major or minor should be evaluated on a case-by-case basis, as determined by the veterinarian and IACUC (NRC 2003b; Silverman et al. 2007; for additional discussion see Chapter 4, Surgical Procedures).

Regardless of classification, multiple surgical procedures on a single animal should be evaluated to determine their impact on the animal's well-being. Multiple major surgical procedures on a single animal are acceptable only if they are (1) included in and essential components of a single research project or protocol, (2) scientifically justified by the investigator, or (3) necessary for clinical reasons. Conservation of scarce animal resources may justify the conduct of multiple major surgeries on a single animal, but the application of such a practice on a single animal used in separate protocols is discouraged and should be reviewed critically by the IACUC. When applicable, the IO must submit a request to the USDA/APHIS and receive approval in order to allow a regulated animal to undergo multiple major survival surgical procedures in separate unrelated research protocols (USDA 1985, 1997a). Justifications for allowing animals not regulated by the USDA to undergo multiple survival procedures that meet the above criteria should conform to those required for regulated species. If multiple survival surgery is approved, the IACUC should pay particular attention to animal well-being through continuing evaluation of outcomes. Cost savings alone is not an adequate reason for performing multiple major survival surgical procedures.

Some procedures characterized as minor may induce substantial post-procedural pain or impairment and should similarly be scientifically justified if performed more than once in a single animal.

Food and Fluid Regulation Regulation of food or fluid intake may be required for the conduct of some physiological, neuroscience, and behavioral research protocols. The regulation process may entail *scheduled access* to food or fluid sources, so an animal consumes as much as desired at regular intervals, or *restriction*, in which the total volume of food or fluid consumed is strictly monitored and controlled (NRC 2003b). The objective when these studies are being planned and executed should be to use the least restriction necessary to achieve the scientific objective while maintaining animal well-being.

The development of animal protocols that involve the use of food or fluid regulation requires the evaluation of three factors: the necessary level of regulation, potential adverse consequences of regulation, and methods for assessing the health and well-being of the animals (NRC 2003b). In addition, the following factors influence the amount of food or fluid restriction that can be safely used in a specific protocol: the species, strain, or stock, gender, and age of the animals; thermoregulatory demand; type of housing; time of feeding, nutritive value, and fiber content of the diet (Heiderstadt et al. 2000; Rowland 2007); and prior experimental manipulation. The degree of food or fluid restriction necessary for consistent behavioral performance is influenced by the difficulty of the task, the individual animal, the motivation required of the animal, and the effectiveness of animal training for a specific protocol-related task.

The animals should be closely monitored to ensure that food and fluid intake meets their nutritional needs (Toth and Gardiner 2000). Body weights should be recorded at least weekly and more often for animals requiring greater restrictions (NRC 2003b). Written records should be maintained for each animal to document daily food and fluid consumption, hydration status, and any behavioral and clinical changes used as criteria for temporary or permanent removal of an animal from a protocol (Morton 2000; NRC 2003b). In the case of conditioned-response research protocols, use of a highly preferred food or fluid as positive reinforcement, instead of restriction, is recommended. Caloric restriction, as a husbandry technique and means of weight control, is discussed in Chapter 3.

Use of Non-Pharmaceutical-Grade Chemicals and Other Substances The use of pharmaceutical-grade chemicals and other substances ensures that toxic or unwanted side effects are not introduced into studies conducted with experimental animals. They should therefore be used, when available, for all animal-related procedures (USDA 1997b). The use of non-pharmaceutical-grade chemicals or substances should be described and justified in the animal use protocol and be approved by the IACUC (Wolff et al. 2003); for example, the use of a non-pharmaceutical-grade chemical or substance may be necessary to meet the scientific goals of a project or when a veterinary or human pharmaceutical-grade product is unavailable. In such instances, consideration should be given to the grade, purity, sterility, pH, pyrogenicity, osmolality, stability, site and route of administration, formulation, compatibility, and pharmacokinetics of the chemical or substance to be administered, as well as animal welfare and scientific issues relating to its use (NIH 2008).

Field Investigations Investigations may involve the observation or use of nondomesticated vertebrate species under field conditions. Many field investigations require international, federal, state, and/or local permits, which may call for an evaluation of the scientific merit of the proposed study and a determination of the potential impact on the population or species to be studied.

Additionally, occupational health and safety issues, including zoonoses, should be reviewed by the institution's health and safety committee or office, with assurances to the IACUC that the field study does not compromise the health and safety of either animals or persons in the field. Principal investigators conducting field research should be knowledgeable about relevant zoonotic diseases, associated safety issues, and any laws or regulations that apply. Exceptions to the above should be clearly defined and evaluated by the IACUC.

In preparing the design of a field study, investigators are encouraged to consult with relevant professional societies and available guidelines (see Appendix A). Veterinary input may be needed for projects involving capture, individual identification, sedation, anesthesia, surgery, recovery, holding, transportation, release, or euthanasia. Issues associated with these activities are similar if not identical to those for species maintained and used in the laboratory. When species are removed from the wild, the protocol should include plans for either a return to their habitat or their final disposition, as appropriate.

The *Guide* does not purport to be a compendium of all information regarding field biology and methods used in wildlife investigations, but the basic principles of humane care and use apply to animals living under natural conditions. IACUCs engaged in the review of field studies are encouraged to consult with a qualified wildlife biologist.

Agricultural Animals The use of agricultural animals in research is subject to the same ethical considerations as for other animals in research, although it is often categorized as either biomedical or agricultural because of government regulations and policies, institutional policies, administrative structure, funding sources, and/or user goals (Stricklin et al. 1990). This categorization has led to a dual system with different criteria for evaluating protocols and standards of housing and care for animals of the same species on the basis of stated biomedical or agricultural research objectives (Stricklin and Mench 1994). With some studies, differences in research goals may lead to a clear distinction between biomedical and agricultural research. For example, animal models of human diseases, organ transplantation, and major surgery are considered biomedical uses; and studies on food and fiber production, such as feeding trials, are usually considered agricultural uses. But when the distinction is unclear, as in the case of some nutrition and disease studies, administrators, regulators, and IACUCs face a dilemma in deciding how to handle such studies (Stricklin et al. 1990). Decisions on categorizing research uses of agricultural animals and defining standards for their care and use should be made by the IACUC based on both the researcher's goals and concern for animal well-being. Regardless of the category of research, institutions are expected to provide oversight of all research animals and ensure that pain and distress are minimized.

The protocol, rather than the category of research, should determine the setting (farm or laboratory). Housing systems for agricultural animals used in biomedical research may or may not differ from those used in agricultural research; animals used in either type of research can be housed in cages, stalls, paddocks, or pastures (Tillman 1994). Some agricultural studies need uniform conditions to minimize environmental variability, and some biomedical studies are conducted in farm settings. Agricultural research often necessitates that animals be managed according to contemporary farm production practices (Stricklin and Mench 1994), and natural environmental conditions might be desirable for agricultural research, whereas control of environmental conditions to minimize variation might be desirable in biomedical research (Tillman 1994).

The *Guide* applies to agricultural animals used in biomedical research, including those maintained in typical farm settings. For animals maintained in a farm setting, the *Guide for the Care and Use of Agricultural Animals in Research and Teaching* (FASS 2010) is a useful resource. Information about environmental enrichment, transport, and handling may be helpful in both agricultural and biomedical research settings. Additional information about facilities and management of farm animals in an agricultural setting is available from the Midwest Plan Service (1987) and from agricultural engineers or animal science experts.

Postapproval Monitoring

Continuing IACUC oversight of animal activities is required by federal laws, regulations, and policies. A variety of mechanisms can be used to facilitate ongoing protocol assessment and regulatory compliance. Postapproval monitoring (PAM) is considered here in the broadest sense, consisting of all types of protocol monitoring after the IACUC's initial protocol approval.

PAM helps ensure the well-being of the animals and may also provide opportunities to refine research procedures. Methods include continuing protocol review; laboratory inspections (conducted either during regular facilities inspections or separately); veterinary or IACUC observation of selected procedures; observation of animals by animal care, veterinary, and IACUC staff and members; and external regulatory inspections and assessments. The IACUC, veterinary, animal care, and compliance staff may all conduct PAM, which may also serve as an educational tool.

Continuing protocol review typically consists of an annual update or review as well as the triennial review required by the PHS. The depth of such reviews varies from a cursory update to a full committee review of the entire protocol. Some institutions use the annual review as an opportunity for the investigator to submit proposed amendments for future procedures, to provide a description of any adverse or unanticipated events, and to provide updates on work progress. For the triennial review, many institutions require a complete new protocol submission and may request a progress report on the use of animals during the previous 3 years.

Both the Health Research Extension Act and the AWA require the IACUC to inspect animal care and use facilities, including sites used for animal surgeries, every 6 months. As part of a formal PAM program some institutions combine inspection of animal study sites with concurrent review of animal protocols. Based on risks to animals and their handlers, other study areas may require more or less frequent inspections. Examples of effective monitoring strategies include

- examination of surgical areas, including anesthetic equipment, use of appropriate aseptic technique, and handling and use of controlled substances
- review of protocol-related health and safety issues
- review of anesthetic and surgical records
- regular review of adverse or unexpected experimental outcomes affecting the animals
- observation of laboratory practices and procedures and comparison with approved protocols.

Institutions may also consider the use of veterinary staff and/or animal health technicians to observe increased risk procedures for adverse events (e.g., novel survival surgeries, pain studies, tumor growth studies) and report their findings for review by the IACUC. The level of formality and intensity of PAM should be tailored to institutional size and complexity, and in all cases should support a culture of care focusing on the animals' well-being (Klein and Bayne 2007). Regardless of the methods used or who conducts and coordinates the monitoring, PAM programs are more likely to succeed when the institution encourages an educational partnership with investigators (Banks and Norton 2008; Collins 2008; Dale 2008; Lowman 2008; Plante and James 2008; Van Sluyters 2008).

DISASTER PLANNING AND EMERGENCY PREPAREDNESS

Animal facilities may be subject to unexpected conditions that result in the catastrophic failure of critical systems or significant personnel absenteeism, or other unexpected events that severely compromise ongoing animal care and well-being (ILAR 2010). Facilities must therefore have a disaster plan. The plan should define the actions necessary to prevent animal pain, distress, and deaths due to loss of systems such as those that control ventilation, cooling, heating, or provision of potable water. If possible the plan should describe how the facility will preserve animals that are necessary for critical research activities or are irreplaceable. Knowledge of the geographic locale may provide guidance as to the probability of a particular type of disaster.

Disaster plans should be established in conjunction with the responsible investigator(s), taking into consideration both the priorities for triaging animal populations and the institutional needs and resources. Animals that cannot be relocated or protected from the consequences of the disaster must be humanely euthanized. The disaster plan should identify essential personnel who should be trained in advance in its implementation. Efforts should be taken to ensure personnel safety and provide access to essential personnel during or immediately after a disaster. Such plans should be approved by the institution and be part of the overall institutional disaster response plan that is coordinated by the IO or another senior-level administrator. Law enforcement and emergency personnel should be provided with a copy of the plan for comment and integration into broader, areawide planning (Vogelweid 1998).

REFERENCES

AAALAC [Association for Assessment and Accreditation of Laboratory Animal Care] International. 2003. Who's responsible for offsite animals? Connection Spring:6-11, 13. Available at www.aaalac.org/publications.

ACLAM [American College of Laboratory Animal Medicine]. 1996. Adequate Veterinary Care. Available at www.aclam.org/education/guidelines/position_adequatecare.html; accessed May 10, 2010.

Anderson LC. 2007. Institutional and IACUC responsibilities for animal care and use education and training programs. ILAR J 48:90-95.

AVMA [American Veterinary Medical Association]. 2008. Introduction to Ergonomics Guidelines for Veterinary Practice. April. Available at www.avma.org/issues/policy/ergonomics.asp; accessed May 10, 2010.

AVMA. 2010. Programs accredited by the AVMA Committee on Veterinary Technician Education and Activities (CVTEA). Available at www.avma.org/education/cvea/vettech_programs/vettech_programs.asp; accessed January 4, 2010.

Banks RE, Norton JN. 2008. A sample postapproval monitoring program in academia. ILAR J 49:402-418.

Bayne KA, Garnett NL. 2008. Mitigating risk, facilitating research. ILAR J 49:369-371.

Boissy A, Manteuffel G, Jensen MB, Moe RO, Spruijt B, Keeling L, Winckler C, Forkman B, Dimitrov I, Langbein J, Bakken M, Veissier I, Aubert A. 2007. Assesment of positive emotions in animals to improve their welfare. Physiol Behav 92:375-397.

Brown RE, Stanford L, Schellinck HM. 2000. Developing standardized behavioral tests for knockout and mutant mice. ILAR J 41:163-174.

Bush RK. 2001. Assessment and treatment of laboratory animal allergy. ILAR J 42:55-64.

Bush RK, Stave GM. 2003. Laboratory animal allergy: An update. ILAR J 44:28-51.

CCAC [Canadian Council on Animal Care]. 1993. Guide to the Care and Use of Experimental Animals, vol 1, 2nd ed. Olfert ED, Cross BM, McWilliam AA, eds. Ontario: CCAC.

CCAC. 1998. Guidelines on Choosing an Appropriate Endpoint in Experiments Using Animals for Research, Teaching and Testing. Ottawa. Available at www.ccac.ca/en/CCAC_Programs/Guidelines_Policies/gdlines/endpts/appopen.htm; accessed May 10, 2010.

CDC [Centers for Disease Control and Prevention] and NIH [National Institutes of Health]. 2000. Primary Containment for Biohazards: Selection, Installation and Use of Biological Safety Cabinets, 2nd ed. Washington: Government Printing Office. Available at www.cdc.gov/od/ohs/biosfty/bsc/bsc.htm; accessed May 25, 2010.

CFR [Code of Federal Regulations]. 1984a. Title 10, Part 20. Standards for Protection against Radiation. Washington: Office of the Federal Register.

CFR. 1984b. Title 29, Part 1910, Occupational Safety and Health Standards, Subpart G, Occupational Health and Environmental Control, and Subpart Z, Toxic and Hazardous Substances. Washington: Office of the Federal Register.

CFR. 1984c. Title 29, Part 1910. Occupational Safety and Health Standards; Subpart I, Personal Protective Equipment. Washington: Office of the Federal Register.

CFR. 1998. Title 29, Section 1910.120. Inspection Procedures for the Hazardous Waste Operations and Emergency Response Standard. Washington: Office of the Federal Register. April 24.

CFR. 2002a. Title 42, Part 73. Possession, Use and Transfer of Select Agents and Toxins. Washington: Office of the Federal Register. December 13.

CFR. 2002b. Title 7, Part 331; and Title 9, Part 121. Agricultural Bioterrorism Protection Act of 2002: Possession, Use and Transfer of Select Agents and Toxins. Washington: Office of the Federal Register. December 13.

Cohen JI, Davenport DS, Stewart JA, Deitchmann S, Hilliard JK, Chapman LE, B Virus Working Group. 2002. Recommendations for prevention of and therapy for exposure to B virus (*Cercopithecine herpesvirus* 1). Clin Infect Dis 35:1191-1203.

Colby LA, Turner PV, Vasbinder MA. 2007. Training strategies for laboratory animal veterinarians: Challenges and opportunities. ILAR J 48:143-155.

Collins JG. 2008. Postapproval monitoring and the IACUC. ILAR J 49:388-392.

Conarello SL, Shepard MJ. 2007. Training strategies for research investigators and technicians. ILAR J 48:120-130.

Crawley JN. 1999. Behavioral phenotyping of transgenic and knockout mice: Experimental design and evaluation of general health, sensory functions, motor abilities, and specific behavioral tests. Brain Res 835:18-26.

Dale WE. 2008. Postapproval monitoring and the role of the compliance office. ILAR J 49:393-401.

Dennis MB. 1999. Institutional animal care and use committee review of genetic engineering. In: Gonder JC, Prentice ED, Russow L-M, eds. Genetic Engineering and Animal Welfare: Preparing for the 21st Century. Greenbelt MD: Scientists Center for Animal Welfare.

Dennis MB. 2000. Humane endpoints for genetically engineered animal models. ILAR J 41:94-98.

DHHS [Department of Health and Human Services]. 2009. Biosafety in Microbiological and Biomedical Laboratories, 5th ed. Chosewood LC, Wilson DE, eds. Washington: Government Printing Office. Available at http://www.cdc.gov/biosafety/publications/bmbl5/index.htm; accessed July 30, 2010.

FASS [Federation of Animal Science Societies]. 2010. Guide for the Care and Use of Agricultural Animals in Research and Teaching, 3rd ed. Champlain, IL: FASS.

Fechter LD. 1995. Combined effects of noise and chemicals. Occup Med 10:609-621.

Foshay WR, Tinkey PT. 2007. Evaluating the effectiveness of training strategies: Performance goals and testing. ILAR J 48:156-162.

Frasier D, Talka J. 2005. Facility design considerations for select agent animal research. ILAR J 46:23-33.

Gonder JC. 2002. Regulatory compliance. In: Suckow MA, Douglas FA, Weichbrod RH, eds. Management of Laboratory Animal Care and Use Programs. Boca Raton, FL: CRC Press. p 163-185.

Gondo Y. 2008. Trends in large-scale mouse mutagenesis: From genetics to functional genomics. Nat Rev Genet 9:803-810.

Gordon S. 2001. Laboratory animal allergy: A British perspective on a global problem. ILAR J 42:37-46.

Gordon S, Wallace J, Cook A, Tee RD, Newman Taylor AJ. 1997. Reduction of exposure to laboratory animal allergens in the workplace. Clin Exp Allergy 27:744-751.

Greene ME, Pitts ME, James ML. 2007. Training strategies for institutional animal care and use committee (IACUC) members and the institutional official (IO). ILAR J 48:131-142.

Harrison DJ. 2001. Controlling exposure to laboratory animal allergens. ILAR J 42:17-36.

Heiderstadt KM, McLaughlin RM, Wright DC, Walker SE, Gomez-Sanchez CE. 2000. The effect of chronic food and water restriction on open-field behaviour and serum corticosterone levels in rats. Lab Anim 34:20-28.

Hendriksen CFM, Steen B. 2000. Refinement of vaccine potency testing with the use of humane endpoints. ILAR J 41:105-113.

Huerkamp MJ, Gladle MA, Mottet MP, Forde K. 2009. Ergonomic considerations and allergen management. In: Hessler JR, Lerner NMD, eds. Planning and Designing Research Animal Facilities. San Diego: Elsevier. p 115-128.

ILAR [Institute for Laboratory Animal Research, National Research Council]. 2000. Humane Endpoints for Animals Used in Biomedical Research and Testing. ILAR J 41:59-123.

ILAR. 2010. Disaster planning and management. ILAR J 51:101-192.

IRAC [Interagency Research Animal Committee]. 1985. US Government Principles for Utilization and Care of Vertebrate Animals Used in Testing, Research, and Training. Federal Register, May 20, 1985. Washington: Office of Science and Technology Policy. Available at http://oacu.od.nih.gov/regs/USGovtPrncpl.htm; accessed May 10, 2010.

Klein HJ, Bayne KA. 2007. Establishing a culture of care, conscience, and responsibility: Addressing the improvement of scientific discovery and animal welfare through science-based performance standards. ILAR J 48:3-11.

Kreger MD. 1995. Training Materials for Animal Facility Personnel: AWIC Quick Bibliography Series, 95-08. Beltsville MD: National Agricultural Library.

Lassnig C, Kolb A, Strobl B, Enjuanes L, Müller M. 2005. Studying human pathogens in human models: Fine tuning the humanized mouse. Transgenic Res 14:803-806.

Laule GE, Bloomsmith MA, Schapiro SJ. 2003. The use of positive reinforcement training techniques to enhance the care, management, and welfare of primates in the laboratory. J Appl Anim Welf Sci 6:163-173.

Lowman RP. 2008. The institutional official and postapproval monitoring: The view from 10,000 feet. ILAR J 49:379-387.

Mann MD, Prentice ED. 2004. Should IACUCs review scientific merit of animal research projects? Lab Anim (NY) 33:26-31.

McCullough NV. 2000. Personal respiratory protection. In: Fleming DO, Hunt DL, eds. Biological Safety Principles and Practices. Washington: ASM Press. p 383-404.

Meunier LD. 2006. Selection, acclimation, training and preparation of dogs for the research setting. ILAR J 47:326-347.

Midwest Plan Service. 1987. Structures and Environment Handbook, 11th ed. rev. Ames: Midwest Plan Service, Iowa State University.

Miller G. 2007. Science and the public: Animal extremists get personal. Science 318:1856-1858.

Moorehead RA, Sanchez OH, Baldwin RM, Khokha R. 2003. Transgenic overexpression of IGF-II induces spontaneous lung tumors: A model for human lung adenocarcinoma. Oncogene 22:853-857.

Morton DB. 2000. A systematic approach for establishing humane endpoints. ILAR J 41:80-86.

Morton WR, Knitter GH, Smith PV, Susor TG, Schmitt K. 1987. Alternatives to chronic restraint of nonhuman primates. JAVMA 191:1282-1286.

Mumphrey SM, Changotra H, Moore TN, Heimann-Nichols ER, Wobus CE, Reilly MJ, Moghadamfalahi M, Shukla D, Karst SM. 2007. Murine norovirus 1 infection is associated with histopathological changes in immunocompetent hosts, but clinical disease is prevented by STAT1-dependent interferon responses. J Virol 81:3251-3263.

Newcomer CE. 2002. Hazard identification and control. In: Suckow MA, Douglas FA, Weichbrod RH, eds. Management of Laboratory Animal Care and Use Programs. Boca Raton, FL: CRC Press. p 291-324.

NIH [National Institutes of Health]. 2002. Guidelines for Research Involving Recombinant DNA Molecules. April. Available at http://oba.od.nih.gov/rdna/nih_guidelines_oba.html; accessed May 20, 2010.

NIH. 2008. Guidelines for the Use of Non-Pharmaceutical-Grade Chemicals/Compounds in Laboratory Animals. Animal Research Advisory Committee, Office of Animal Care and Use, NIH. Available at http://oacu.od.nih.gov/ARAC/documents/Pharmaceutical_Compounds.pdf; accessed May 20, 2010.

NIOSH [National Institute for Occupational Safety and Health]. 1997a. Elements of Ergonomics Programs: A Primer Based on Workplace Evaluations of Musculoskeletal Disorders (NIOSH Publication No. 97-117). Cincinnati: NIOSH. p 16-24.

NIOSH. 1997b. Musculoskeletal Disorders and Workplace Factors: A Critical Review of Epidemiologic Evidence for Work-Related Musculoskeletal Disorders of the Neck, Upper Extremity, and Low Back. Bernard B, ed. Cincinnati: DHHS, PHS, CDDC, NIOSH. p 1-12.

NRC [National Research Council]. 1991. Education and Training in the Care and Use of Laboratory Animals: A Guide for Developing Institutional Programs. Washington: National Academy Press.

NRC. 1997. Occupational Health and Safety in the Care and Use of Research Animals. Washington: National Academy Press.

NRC. 2003a. Occupational Health and Safety in the Care and Use of Nonhuman Primates. Washington: National Academies Press.

NRC. 2003b. Guidelines for the Care and Use of Mammals in Neuroscience and Behavioral Research. Washington: National Academies Press.

NRC. 2004. Biotechnology Research in an Age of Terrorism. Washington: National Academies Press.

OECD [Organisation for Economic Co-operation and Development]. 1999. Guidance Document on Humane Endpoints for Experimental Animals Used in Safety Evaluation Studies. Paris: OECD.

Olfert ED, Godson DL. 2000. Humane endpoints for infectious disease animal models. ILAR J 41:99-104.
OSHA [Occupational Safety and Health Administration]. 1998a. Occupational Safety and Health Standards. Subpart G, Occupational Health and Environmental Controls (29 CFR 1910). Washington: Department of Labor.
OSHA. 1998b. Occupational Safety and Health Standards. Subpart Z, Toxic and Hazardous Substances, Bloodborne Pathogens (29 CFR 1910.1030). Washington: Department of Labor.
OSHA. 1998c. Occupational Safety and Health Standards. Subpart G, Occupational Health and Environmental Controls, Occupational Noise Exposure (29 CFR 1910.95). Washington: Department of Labor.
OSHA. 1998d. Occupational Safety and Health Standards. Subpart I, Personal Protective Equipment, Respiratory Protection (29 CFR 1910.134). Washington: Department of Labor.
Paster EV, Villines KA, Hickman DL. 2009. Endpoints for mouse abdominal tumor models: Refinement of current criteria. Comp Med 59:234-241.
PHS [Public Health Service]. 2002. Public Health Service Policy on Humane Care and Use of Laboratory Animals. Department of Health and Human Services, National Institutes of Health, Office of Laboratory Animal Welfare. Available at http://grants.nih.gov/grants/olaw/references/phspol.htm; accessed January 14, 2010.
PL [Public Law] 104-191. 1996. Health Insurance Portability and Accountability Act (HIPAA) of 1996. Washington: Government Printing Office.
PL 107-56. 2001. Uniting and Strengthening America by Providing Appropriate Tools Required to Intercept and Obstruct Terrorism (USA PATRIOT) Act of 2001. Washington: Government Printing Office. October 26.
PL 107-188. 2002. Public Health Security and Bioterrorism Preparedness and Response Act of 2002. Washington: Government Printing Office. June 12.
Plante A, James ML. 2008. Program oversight enhancements (POE): The big PAM. ILAR J 49:419-425.
Prescott MJ, Buchanan-Smith HM. 2003. Training nonhuman primates using positive reinforcement techniques. J Appl Anim Welf Sci 6:157-161.
Pritt S, Duffee N. 2007. Training strategies for animal care technicians and veterinary technical staff. ILAR J 48:109-119.
Reeb-Whitaker CK, Harrison, DJ, Jones RB, Kacergis JB, Myers DD, Paigen B. 1999. Control strategies for aeroallergens in an animal facility. J Allergy Clin Immunol 103:139-146.
Reinhardt V. 1991. Training adult male rhesus monkeys to actively cooperate during in-homecage venipuncture. Anim Technol 42:11-17.
Reinhardt V. 1995. Restraint methods of laboratory non-human primates: A critical review. Anim Welf 4:221-238.
Richmond JY, Hill RH, Weyant RS, Nesby-O'Dell SL, Vinson PE. 2003. What's hot in animal biosafety? ILAR J 44:20-27.
Rowland NE. 2007. Food or fluid restriction in common laboratory animals: Balancing welfare considerations with scientific inquiry. Comp Med 57:149-160.
Sargent EV, Gallo F. 2003. Use of personal protective equipment for respiratory protection. ILAR J 44:52-56.
Sass N. 2000. Humane endpoints and acute toxicity testing. ILAR J 41:114-123.
Sauceda R, Schmidt MG. 2000. Refining macaque handling and restraint techniques. Lab Anim 29:47-49.
Schweitzer IB, Smith E, Harrison DJ, Myers DD, Eggleston PA, Stockwell JD, Paigen B, Smith AL. 2003. Reducing exposure to laboratory animal allergens. Comp Med 53:487-492.
Seward JP. 2001. Medical surveillance of allergy in laboratory animal handlers. ILAR J 42:47-54.
Silverman J, Sukow MA, Murthy S, eds. 2007. The IACUC Handbook, 2nd ed. Boca Raton, FL: CRC Press.

Stokes WS. 2000. Reducing unrelieved pain and distress in laboratory animals using humane endpoints. ILAR J 41:59-61.

Stokes WS. 2002. Humane endpoints for laboratory animals used in regulatory testing. ILAR J 43:S31-S38.

Stricklin WR, Mench JA. 1994. Oversight of the use of agricultural animals in university teaching and research. ILAR News 36:9-14.

Stricklin WR, Purcell D, Mench JA. 1990. Farm animals in agricultural and biomedical research. In: The Well-Being of Agricultural Animals in Biomedical and Agricultural Research: Proceedings from a SCAW-Sponsored Conference, September 6-7. Washington: Scientists Center for Animal Welfare. p 1-4.

Thomann WR. 2003. Chemical safety in animal care, use, and research. ILAR J 44:13-19.

Thulin H, Bjorkdahl M, Karlsson AS, Renstrom A. 2002. Reduction of exposure to laboratory animal allergens in a research laboratory. Ann Occup Hyg 46:61-68.

Tillman P. 1994. Integrating agricultural and biomedical research policies: Conflicts and opportunities. ILAR News 36:29-35.

Toth LA. 1997. The moribund state as an experimental endpoint. Contemp Top Lab Anim Sci 36:44-48.

Toth LA. 2000. Defining the moribund condition as an experimental endpoint for animal research. ILAR J 41:72-79.

Toth LA, Gardiner TW. 2000. Food and water restriction protocols: Physiological and behavioral considerations. Contemp Top Lab Anim Sci 39:9-17.

UKCCCR [United Kingdom Coordinating Committee on Cancer Research]. 1997. Guidelines for the Welfare of Animals in Experimental Neoplasia, 2nd ed. London: UKCCCR.

USC [United States Code]. Title 42, Chapter 6a, Subchapter III, Part H, Section 289d: Animals in Research. Available at http://uscode.house.gov/download/pls/42CGA.txt.

USDA [US Department of Agriculture]. 1985. 9 CFR 1A. (Title 9, Chapter 1, Subchapter A): Animal Welfare. Available at http://ecfr.gpoaccess.gov/cgi/t/text/text-idx?sid=8314313bd7adf2c9f1964e2d82a88d92andc=ecfrandtpl=/ecfrbrowse/Title09/9cfrv1_02.tpl; accessed January 14, 2010.

USDA. 1997a. APHIS Policy #14, "Multiple Survival Surgery: Single vs. Multiple Procedures" (April 14). Available at www.aphis.usda.gov/animal_welfare/downloads/policy/policy14.pdf; accessed January 4, 2010.

USDA. 1997b. APHIS Policy #3, "Veterinary Care" (July 17). Available at www.aphis.usda.gov/animal_welfare/downloads/policy/policy3.pdf; accessed January 9, 2010.

USDA. 2002. Facilities Design Standards. Manual 242.1. Available at www.afm.ars.usda.gov/ppweb/PDF/242-01M.pdf; accessed May 10, 2010.

Van Sluyters RC. 2008. A guide to risk assessment in animal care and use programs: The metaphor of the 3-legged stool. ILAR J 49:372-378.

Yeates JW, Main DCJ. 2009. Assesment of positive welfare: A review. Vet Rev 175:293-300.

Vogelweid CM. 1998. Developing emergency management plans for university laboratory animal programs and facilities. Contemp Top Lab Anim Sci 37:52-56.

Wallace J. 2000. Humane endpoints and cancer research. ILAR J 41:87-93.

Wolff A, Garnett N, Potkay S, Wigglesworth C, Doyle D, Thornton, D. 2003. Frequently asked questions about the Public Health Service Policy on Humane Care and Use of Laboratory Animals. Lab Anim 32(9):33-36.

Wolfle TL, Bush RK. 2001. The science and pervasiveness of laboratory animal allergy. ILAR J 42:1-3.

Wood RA. 2001. Laboratory animal allergens. ILAR J 42:12-16.

3

Environment, Housing, and Management

This chapter provides guidelines for the environment, housing, and management of laboratory animals used or produced for research, testing, and teaching. These guidelines are applicable across species and are relatively general; additional information should be sought about how to apply them to meet the specific needs of any species, strain, or use (see Appendix A for references). The chapter is divided into recommendations for terrestrial (page 42) and aquatic animals (page 77), as there are fundamental differences in their environmental requirements as well as animal husbandry, housing, and care needs. Although formulated specifically for vertebrate species, the general principles of humane animal care as set out in the *Guide* may also apply to invertebrate species.

The design of animal facilities combined with appropriate animal housing and management are essential contributors to animal well-being, the quality of animal research and production, teaching or testing programs involving animals, and the health and safety of personnel. An appropriate Program (see Chapter 2) provides environments, housing, and management that are well suited for the species or strains of animals maintained and takes into account their physical, physiologic, and behavioral needs, allowing them to grow, mature, and reproduce normally while providing for their health and well-being.

Fish, amphibians, and reptiles are *poikilothermic* animals: their core temperature varies with environmental conditions and they have limited ability (compared with birds and mammals) to metabolically maintain core temperature. The majority of poikilothermic laboratory animals are aquatic species—for example, fish and most amphibians—although some, such as reptiles and certain amphibian species, are terrestrial. Personnel working with aquatic animals should be familiar with management implications, e.g., the importance of providing appropriate temperature ranges for basic physiologic function.

TERRESTRIAL ANIMALS

Terrestrial Environment

Microenvironment and Macroenvironment

The *microenvironment* of a terrestrial animal is the physical environment immediately surrounding it; that is, the primary enclosure such as the cage, pen, or stall. It contains all the resources with which the animals come directly in contact and also provides the limits of the animals' immediate environment. The microenvironment is characterized by many factors, including illumination, noise, vibration, temperature, humidity, and gaseous and particulate composition of the air. The physical environment of the secondary enclosure, such as a room, a barn, or an outdoor habitat, constitutes the *macroenvironment*.

> *Microenvironment:* The immediate physical environment surrounding the animal (i.e., the environment in the primary enclosure such as the cage, pen, or stall).

Although the microenvironment and the macroenvironment are generally related, the microenvironment can be appreciably different and affected by several factors, including the design of the primary enclosure and macroenvironmental conditions.

Evaluation of the microenvironment of small enclosures can be difficult. Available data indicate that temperature, humidity, and concentrations of gases and particulate matter are often higher in the animal microenvironment than in the macroenvironment (Besch 1980; Hasenau et al. 1993; Perkins and Lipman 1995; E. Smith et al. 2004), while light levels are usually lower. Microenvironmental conditions can directly affect physiologic processes and behavior and may alter disease susceptibility (Baer et al. 1997; Broderson et al. 1976; Memarzadeh et al. 2004; Schoeb et al. 1982; Vesell et al. 1976).

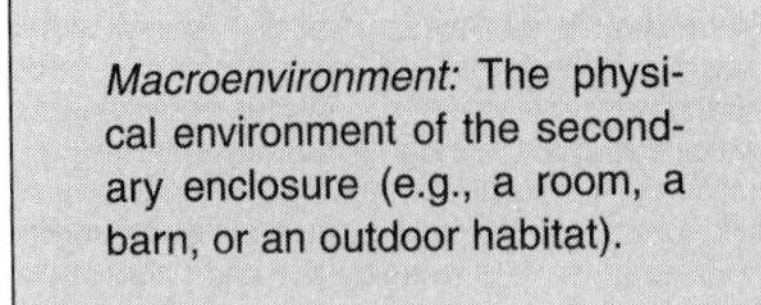
Macroenvironment: The physical environment of the secondary enclosure (e.g., a room, a barn, or an outdoor habitat).

Temperature and Humidity

Maintenance of body temperature within normal circadian variation is necessary for animal well-being. Animals should be housed within temperature and humidity ranges appropriate for the species, to which they can adapt with minimal stress and physiologic alteration.

The ambient temperature range in which thermoregulation occurs without the need to increase metabolic heat production or activate evaporative heat loss mechanisms is called the *thermoneutral zone (TNZ)* and is bounded by the lower and upper critical temperatures (LCTs and UCTs; Gordon 2005). To maintain body temperature under a given environmental temperature animals adjust physiologically (including their metabolism) and behaviorally (including their activity level and resource use). For example, the TNZ of mice ranges between 26°C and 34°C (Gordon 1993); at lower temperatures, building nests and huddling for resting and sleeping allow them to thermoregulate by behaviorally controlling their microclimate. Although mice choose temperatures below their LCT of 26°C during activity periods, they strongly prefer temperatures above their LCT for maintenance and resting behaviors (Gaskill et al. 2009; Gordon 2004; Gordon et al. 1998). Similar LCT values are found in the literature for other rodents, varying between 26-30°C for rats and 28-32°C for gerbils (Gordon 1993). The LCTs of rabbits (15-20°C; Gonzalez et al. 1971) and cats and dogs (20-25°C) are slightly lower, while those of nonhuman primates and farm animals vary depending on the species. In general, dry-bulb temperatures in animal rooms should be set below the animals' LCT to avoid heat stress. This, in turn, means that animals should be provided with adequate resources for thermoregulation (nesting material, shelter) to avoid cold stress. Adequate resources for thermoregulation are particularly important for newborn animals whose LCT is normally considerably higher than that of their adult conspecifics.

Environmental temperature and relative humidity can be affected by husbandry and housing design and can differ considerably between primary and secondary enclosures as well as within primary enclosures. Factors that contribute to variation in temperature and humidity between and within enclosures include housing design; construction material; enrichment devices such as shelters and nesting material; use of filter tops; number, age, type, and size of the animals in each enclosure; forced ventilation of enclosures; and the type and frequency of contact bedding changes (Besch 1980).

Exposure to wide temperature and humidity fluctuations or extremes may result in behavioral, physiologic, and morphologic changes, which might negatively affect animal well-being and research performance as well as outcomes of research protocols (Garrard et al. 1974; Gordon 1990,

1993; Pennycuik 1967). These effects can be multigenerational (Barnett 1965, 1973).

The dry-bulb temperatures listed in Table 3.1 are broad and generally reflect tolerable limits for common adult laboratory animal species, provided they are housed with adequate resources for behavioral thermoregulation; temperatures should normally be selected and maintained with minimal fluctuation near the middle of these ranges. Depending on the specific housing system employed, the selection of appropriate macro- and microenvironmental temperatures will differ based on a variety of factors, including but not limited to the species or strain, age, numbers of animals in the enclosure, size and construction of the primary enclosure, and husbandry conditions (e.g., use/provision of contact bedding, nesting material and/or shelter, individually ventilated cages). Poikilotherms and young birds of some species generally require a thermal gradient in their primary enclosure to meet basic physiological processes. The temperature ranges shown may not apply to captive wild animals, wild animals maintained in their natural environment, or animals in outdoor enclosures that have the opportunity to adapt by being exposed to seasonal changes in ambient conditions.

Some conditions require increased environmental temperatures for housing (e.g., postoperative recovery, neonatal animals, rodents with hairless phenotypes, reptiles and amphibians at certain stages of reproduction). The magnitude of the temperature increase depends on housing details; sometimes raising the temperature in the microenvironment alone (e.g., by using heating pads for postoperative recovery or radiant heat sources for reptiles) rather than raising the temperature of the macroenvironment is sufficient and preferable.

Relative humidity should also be controlled, but not nearly as narrowly as temperature for many mammals; the acceptable range of relative humidity is considered to be 30% to 70% for most mammalian species. Micro-

TABLE 3.1 Recommended Dry-Bulb Macroenvironmental Temperatures for Common Laboratory Animals

	Dry-Bulb Temperature	
Animal	°C	°F
Mouse, rat, hamster, gerbil, guinea pig[a]	20-26	68-79
Rabbit	16-22	61-72
Cat, dog, nonhuman primate	18-29	64-84
Farm animals, poultry	16-27	61-81

[a]Dry-bulb room temperature settings for rodents are typically set below the animals' LCT to avoid heat stress, and should reflect different species-specific LCT values. Animals should be provided with adequate resources for thermoregulation (nesting material, shelter) to avoid cold stress.

environmental relative humidity may be of greater importance for animals housed in a primary enclosure in which the environmental conditions differ greatly from those of the macroenvironment (e.g., in static filter-top [isolator] cages).

Some species may require conditions with high relative humidity (e.g., selected species of nonhuman primates, tropical reptiles, and amphibians; Olson and Palotay 1983). In mice, both abnormally high and low humidity may increase preweaning mortality (Clough 1982). In rats, low relative humidity, especially in combination with temperature extremes, may lead to ringtail, a condition involving ischemic necrosis of the tail and sometimes toes (Crippa et al. 2000; Njaa et al. 1957; Totten 1958). For some species, elevated relative humidity may affect an animal's ability to cope with thermal extremes. Elevated microenvironmental relative humidity in rodent isolator cages may also lead to high intracage ammonia concentrations (Corning and Lipman 1991; Hasenau et al. 1993), which can be irritating to the nasal passages and alter some biologic responses (Gordon et al. 1980; Manninen et al. 1998). In climates where it is difficult to provide a sufficient level of environmental relative humidity, animals should be closely monitored for negative effects such as excessively flaky skin, ecdysis (molting) difficulties in reptiles, and desiccation stress in semiaquatic amphibians.

Ventilation and Air Quality

The primary purpose of ventilation is to provide appropriate air quality and a stable environment. Specifically, ventilation provides an adequate oxygen supply; removes thermal loads caused by the animals, personnel, lights, and equipment; dilutes gaseous and particulate contaminants including allergens and airborne pathogens; adjusts the moisture content and temperature of room air; and, where appropriate, creates air pressure differentials (directional air flow) between adjoining spaces. Importantly, ventilating the room (i.e., the macroenvironment) does not necessarily ensure adequate ventilation of an animal's primary enclosure (i.e., the microenvironment), that is, the air to which the animal is actually exposed. The type of primary enclosure may considerably influence the differences between these two environments—for example, differences may be negligible when animals are housed in open caging or pens, whereas they can be significant when static isolator cages are used.

The volume and physical characteristics of the air supplied to a room and its diffusion pattern influence the ventilation of an animal's primary enclosure and are important determinants of the animal's microenvironment. The type and location of supply air diffusers and exhaust registers in relation to the number, arrangement, location, and type of primary and secondary enclosures affect how well the microenvironments are ventilated and should therefore be considered. The use of computer modeling for assessing those factors in relation to heat loading, air diffusion patterns, and particulate movement may be helpful in optimizing ventilation of micro- and macroenvironments (Hughes and Reynolds 1995).

Direct exposure of animals to air moving at high velocity (drafts) should be avoided as the speed of air to which animals are exposed affects the rate at which heat and moisture are removed from an animal. For example, air at 20°C moving at 60 linear feet per minute (18.3 m/min) has a cooling effect of approximately 7°C (Weihe 1971). Drafts can be particularly problematic for neonatal homeotherms (which may be hairless and have poorly developed mechanisms for thermoregulatory control), for mutants lacking fur, and for semiaquatic amphibians that can desiccate.

Provision of 10 to 15 fresh air changes per hour in animal housing rooms is an acceptable guideline to maintain macroenvironmental air quality by constant volume systems and may also ensure microenvironmental air quality. Although this range is effective in many animal housing settings, it does not take into account the range of possible heat loads; the species, size, and number of animals involved; the type of primary enclosure and bedding; the frequency of cage changing; the room dimensions; or the efficiency of air distribution both in the macroenvironment and between the macro- and microenvironments. In some situations, the use of such a broad guideline might overventilate a macroenvironment containing few animals, thereby wasting energy, or underventilate a microenvironment containing many animals, allowing heat, moisture, and pollutants to accumulate.

Modern heating, ventilation, and air conditioning (HVAC) systems (e.g., variable air volume, or VAV, systems) allow ventilation rates to be set in accordance with heat load and other variables. These systems offer considerable advantages with respect to flexibility and energy conservation, but should always provide a minimum amount of air exchange, as recommended for general use laboratories (Bell 2008; DiBerardinis et al. 2009).

Individually ventilated cages (IVCs) and other types of specialized primary enclosures, that either directly ventilate the enclosure using filtered room air or are ventilated independently of the room, can effectively address animals' ventilation requirements without the need to increase macroenvironmental ventilation. However, cautions mentioned above regarding high-velocity air should be considered (Baumans et al. 2002; Krohn et al. 2003). Nevertheless, the macroenvironment should be ventilated sufficiently to address heat loads, particulates, odors, and waste gases released from primary enclosures (Lipman 1993).

If ventilated primary enclosures have adequate filtration to address contamination risks, air exhausted from the microenvironment may be returned to the room in which animals are housed, although it is generally preferable to exhaust these systems directly into the building's exhaust system to reduce heat load and macroenvironmental contamination.

Static isolation caging (without forced ventilation), such as that used in some types of rodent housing, restricts ventilation (Keller et al. 1989). To compensate, it may be necessary to adjust husbandry practices, including sanitation and cage change frequency, selection of contact bedding, placement of cages in a secondary enclosure, animal densities in cages, and/or decrease in macroenvironmental relative humidity to improve the microenvironment and heat dissipation.

The use of recycled air to ventilate animal rooms may save energy but entails risks. Because many animal pathogens can be airborne or travel on fomites (e.g., dust), exhaust air recycled into HVAC systems that serve multiple rooms presents a risk of cross contamination. Recycling air from nonanimal use areas (e.g., some human occupancy areas and food, bedding, and supply storage areas) may require less intensive filtration or conditioning and pose less risk of infection. The risks in some situations, however, might be too great to consider recycling (e.g., in the case of nonhuman primates and biohazard areas). The exhaust air to be recycled should be filtered, at minimum, with 85-95% ASHRAE efficient filters to remove airborne particles before it is recycled (NAFA 1996). Depending on the air source, composition, and proportion of recycled air used (e.g., ammonia and other gases emitted from excrement in recirculating air from animal rooms), consideration should also be given to filtering volatile substances. In areas that require filtration to ensure personnel and/or animal safety (e.g., hazardous containment holding), filter efficiency, loading, and integrity should be assessed.

The successful operation of any HVAC system requires regular preventive maintenance and evaluation, including measurement of its function at the level of the secondary enclosure. Such measurements should include supply and exhaust air volumes, fluctuation in temperature and relative humidity, and air pressure differentials between spaces as well as critical mechanical operating parameters.

Illumination

Light can affect the physiology, morphology, and behavior of various animals (Azar et al. 2008; Brainard et al. 1986; Erkert and Grober 1986; Newbold et al. 1991; Tucker et al. 1984). Potential photostressors include inappropriate photoperiod, photointensity, and spectral quality of the light (Stoskopf 1983).

Numerous factors can affect animals' needs for light and should be considered when an appropriate illumination level is being established for an animal holding room. These include light intensity and wavelength as well as the duration of the animal's current and prior exposure to light, and the animal's pigmentation, circadian rhythm, body temperature, hormonal status, age, species, sex, and stock or strain (Brainard 1989; Duncan and O'Steen 1985; O'Steen 1980; Saltarelli and Coppola 1979; Semple-Rowland and Dawson 1987; Wax 1977). More recent studies in rodents and primates have shown the importance of intrinsically photosensitive retinal ganglion cells (distinct from rods and cones) for neuroendocrine, circadian, and neurobehavioral regulation (Berson et al. 2002; Hanifin and Brainard 2007). These cells can respond to light wavelengths that may differ from other photoreceptors and may influence the type of lighting, light intensity, and wavelength selected for certain types of research.

In general, lighting should be diffused throughout an animal holding area and provide sufficient illumination for the animals' well-being while permitting good housekeeping practices, adequate animal inspection including for the bottom-most cages in racks, and safe working conditions for personnel. Light in animal holding rooms should provide for both adequate vision and neuroendocrine regulation of diurnal and circadian cycles (Brainard 1989).

Photoperiod is a critical regulator of reproductive behavior in many animal species (Brainard et al. 1986; Cherry 1987), so inadvertent light exposure during the dark cycle should be minimized or avoided. Because some species, such as chickens (Apeldoorn et al. 1999), will not eat in low light or darkness, such illumination schedules should be limited to a duration that will not compromise their well-being. A time-controlled lighting system should be used to ensure a regular diurnal cycle, and system performance should be checked regularly to ensure proper cycling.

Most commonly used laboratory rodents are nocturnal. Because albino rodents are more susceptible to phototoxic retinopathy than other animals (Beaumont 2002), they have been used as a basis for establishing room illumination levels (Lanum 1979). Data for room light intensities for other animals, based on scientific studies, are not available. Light levels of about 325 lux (30-ft candles) approximately 1 m (3.3 ft) above the floor appear to be sufficient for animal care and do not cause clinical signs of phototoxic retinopathy in albino rats (Bellhorn 1980). Levels up to 400 lux (37-ft candles) as measured in an empty room 1 m from the floor have been found to be satisfactory for rodents if management practices are used to prevent retinal damage in albinos (Clough 1982). However, the light experience of an individual animal can affect its sensitivity to phototoxicity; light of 130-270 lux above the light intensity under which it was raised has been reported to be near the threshold of retinal damage in some individual albino rats according to histologic, morphometric, and electrophysiologic evidence (Semple-Rowland and Dawson 1987). Some guidelines recommend a light intensity as low as 40 lux at the position of the animal in midcage (NASA 1988). Rats and mice generally prefer cages with low light intensity (Blom et al. 1996), and albino rats prefer areas with a light intensity of less than 25 lux (Schlingmann et al. 1993a). Young mice prefer much lower illumination than adults (Wax 1977). For animals that have been shown to be susceptible to phototoxic retinopathy, light should be between 130 and 325 lux in the room at cage level.

Light intensity decreases with the square of the distance from its source. Thus the location of a cage on a rack affects the intensity of light to which the animals within are exposed. Light intensity may differ as much as 80-fold in transparent cages from the top to the bottom of a rack, and differences up to 20-fold have been recorded within a cage (Schlingmann et al. 1993a,b). Management practices, such as rotating cage position relative to the light source (Greenman et al. 1982) or providing animals with ways to control their own light exposure by behavioral means (e.g., nesting or bedding material adequate for tunneling), can reduce inappropriate light stimulation. Variable-intensity lights are often used to accommodate the needs of research protocols, certain animal species, and energy conservation. However, such a system should also provide for the observation and care of the animals. Caution should be exercised as increases in daytime room illumination for maintenance purposes have been shown to change photoreceptor physiology and can alter circadian regulation (NRC 1996; Reme et al. 1991; Terman et al. 1991).

Noise and Vibration

Noise produced by animals and animal care activities is inherent in the operation of an animal facility (Pfaff and Stecker 1976) and noise control should be considered in facility design and operation (Pekrul 1991). Assessment of the potential effects of noise on an animal warrants consideration of the intensity, frequency, rapidity of onset, duration, and vibration potential of the sound and the hearing range, noise exposure history, and sound effect susceptibility of the species, stock, or strain. Similarly, occupational exposure to animal or animal care practices that generate noise may be of concern for personnel and, if of sufficient intensity, may warrant hearing protection.

Separation of human and animal areas minimizes disturbances to both human and animal occupants of the facility. Noisy animals, such as dogs, swine, goats, nonhuman primates, and some birds (e.g., zebra finches), should be housed away from quieter animals, such as rodents, rabbits, and cats. Environments should be designed to accommodate animals that make noise rather than resorting to methods of noise reduction. Exposure to sound louder than 85 dB can have both auditory and nonauditory effects (Fletcher 1976; Peterson 1980)—for example, eosinopenia, increased adrenal gland weights, and reduced fertility in rodents (Geber et al. 1966; Nayfield and

Besch 1981; Rasmussen et al. 2009), and increased blood pressure in nonhuman primates (Peterson et al. 1981)—and may necessitate hearing protection for personnel (OSHA 1998). Many species can hear sound frequencies inaudible to humans (Brown and Pye 1975; Heffner and Heffner 2007); rodents, for example, are very sensitive to ultrasound (Olivier et al. 1994). The potential effects of equipment (such as video display terminals; Sales 1991; Sales et al. 1999) and materials that produce noise in the hearing range of nearby animals can thus become an uncontrolled variable for research experiments and should therefore be carefully considered (Turner et al. 2007; Willott 2007). To the greatest extent possible, activities that generate noise should be conducted in rooms or areas separate from those used for animal housing.

Because changes in patterns of sound exposure have different effects on different animals (Armario et al. 1985; Clough 1982), personnel should try to minimize the production of unnecessary noise. Excessive and intermittent noise can be minimized by training personnel in alternatives to noisy practices, the use of cushioned casters and bumpers on carts, trucks, and racks, and proper equipment maintenance (e.g., castor lubrication). Radios, alarms, and other sound generators should not be used in animal rooms unless they are part of an approved protocol or enrichment program. Any radios or sound generators used should be switched off at the end of the working day to minimize associated adverse physiologic changes (Baldwin 2007).

While some vibration is inherent to every facility and animal housing condition, excessive vibration has been associated with biochemical and reproductive changes in laboratory animals (Briese et al. 1984; Carman et al. 2007) and can become an uncontrolled variable for research experiments. The source of vibrations may be located within or outside the animal facility. In the latter case, groundborne vibration may affect both the structure and its contents, including animal racks and cages. Housing systems with moving components, such as ventilated caging system blowers, may create vibrations that could affect the animals housed within, especially if not functioning properly. Like noise, vibration varies with intensity, frequency, and duration. A variety of techniques may be used to isolate groundborne (see Chapter 5) and equipment-generated vibration (Carman et al. 2007). Attempts should be made to minimize the generation of vibration, including from humans, and excessive vibration should be avoided.

Terrestrial Housing

Microenvironment (Primary Enclosure)

All animals should be housed under conditions that provide sufficient space as well as supplementary structures and resources required to meet physical, physiologic, and behavioral needs. Environments that fail to meet the animals' needs may result in abnormal brain development, physiologic dysfunction, and behavioral disorders (Garner 2005; van Praag et al. 2000; Würbel 2001) that may compromise both animal well-being and scientific validity. The primary enclosure or space may need to be enriched to prevent such effects (see also section on Environmental Enrichment).

An appropriate housing space or enclosure should also account for the animals' social needs. Social animals should be housed in stable pairs or groups of compatible individuals unless they must be housed alone for experimental reasons or because of social incompatibility (see also section on Behavioral and Social Management). Structural adjustments are frequently required for social housing (e.g., perches, visual barriers, refuges), and important resources (e.g., food, water, and shelter) should be provided in such a way that they cannot be monopolized by dominant animals (see also section on Environmental Enrichment).

The primary enclosure should provide a secure environment that does not permit animal escape and should be made of durable, nontoxic materials that resist corrosion, withstand the rigors of cleaning and regular handling, and are not detrimental to the health and research use of the animals. The enclosure should be designed and manufactured to prevent accidental entrapment of animals or their appendages and should be free of sharp edges or projections that could cause injury to the animals or personnel. It should have smooth, impervious surfaces with minimal ledges, angles, corners, and overlapping surfaces so that accumulation of dirt, debris, and moisture is minimized and cleaning and disinfecting are not impaired. All enclosures should be kept in good repair to prevent escape of or injury to animals, promote physical comfort, and facilitate sanitation and servicing. Rusting or oxidized equipment, which threatens the health or safety of animals, needs to be repaired or replaced. Less durable materials, such as wood, may be appropriate in select situations, such as outdoor corrals, perches, climbing structures, resting areas, and perimeter fences for primary enclosures. Wooden items may need to be replaced periodically because of damage or difficulties with sanitation. Painting or sealing wood surfaces with nontoxic materials may improve durability in many instances.

Flooring should be solid, perforated, or slatted with a slip-resistant surface. In the case of perforated or slatted floors, the holes and slats should have smooth edges. Their size and spacing need to be commensurate with the size of the housed animal to minimize injury and the development of foot lesions. If wire-mesh flooring is used, a solid resting area may be beneficial, as this floor type can induce foot lesions in rodents and rabbits (Drescher 1993; Fullerton and Gilliatt 1967; Rommers and Meijerhof 1996). The size and weight of the animal as well as the duration of housing on wire-mesh floors may also play a role in the development of this condition (Peace et al. 2001). When given the choice, rodents prefer solid floors (with bedding) to grid or wire-mesh flooring (Blom et al. 1996; Manser et al. 1995, 1996).

Animals should have adequate bedding substrate and/or structures for resting and sleeping. For many animals (e.g., rodents) contact bedding expands the opportunities for species-typical behavior such as foraging, digging, burrowing, and nest building (Armstrong et al. 1998; Ivy et al. 2008). Moreover, it absorbs urine and feces to facilitate cleaning and sanitation. If provided in sufficient quantity to allow nest building or burrowing, bedding also facilitates thermoregulation (Gordon 2004). Breeding animals should have adequate nesting materials and/or substitute structures based on species-specific requirements (mice: Sherwin 2002; rats: Lawlor 2002; gerbils: Waiblinger 2002).

Specialized housing systems (e.g., isolation-type cages, IVCs, and *gnotobiotic*[1] isolators) are available for rodents and certain species. These systems, designed to minimize the spread of airborne particles between cages or groups of cages, may require different husbandry practices, such as alterations in the frequency of bedding change, the use of aseptic handling techniques, and specialized cleaning, disinfecting, or sterilization regimens to prevent microbial transmission by other than airborne routes.

Appropriate housing strategies for a particular species should be developed and implemented by the animal care management, in consultation with the animal user and veterinarian, and reviewed by the IACUC. Housing should provide for the animals' health and well-being while being consistent with the intended objectives of animal use. Expert advice should be sought when new species are housed or when there are special requirements associated with the animals or their intended use (e.g., genetically modified animals, invasive procedures, or hazardous agents). Objective assessments should be made to substantiate the adequacy of the animal's environment, housing, and management. Whenever possible, routine procedures for maintaining animals should be documented to ensure consistency of management and care.

Environmental Enrichment

The primary aim of environmental enrichment is to enhance animal well-being by providing animals with sensory and motor stimulation, through structures and resources that facilitate the expression of species-typical behaviors and promote psychological well-being through physical

[1]Gnotobiotic: germ-free animals or formerly germ-free animals in which the composition of any associated microbial flora, if present, is fully defined (Stedman's Electronic Medical Dictionary 2006. Lippincott Williams & Wilkins).

exercise, manipulative activities, and cognitive challenges according to species-specific characteristics (NRC 1998a; Young 2003). Examples of enrichment include structural additions such as perches and visual barriers for nonhuman primates (Novak et al. 2007); elevated shelves for cats (Overall and Dyer 2005; van den Bos and de Cock Buning 1994) and rabbits (Stauffacher 1992); and shelters for guinea pigs (Baumans 2005), as well as manipulable resources such as novel objects and foraging devices for nonhuman primates; manipulable toys for nonhuman primates, dogs, cats, and swine; wooden chew sticks for some rodent species; and nesting material for mice (Gaskill et al. 2009; Hess et al. 2008; Hubrecht 1993; Lutz and Novak 2005; Olsson and Dahlborn 2002). Novelty of enrichment through rotation or replacement of items should be a consideration; however, changing animals' environment too frequently may be stressful.

Well-conceived enrichment provides animals with choices and a degree of control over their environment, which allows them to better cope with environmental stressors (Newberry 1995). For example, visual barriers allow nonhuman primates to avoid social conflict; elevated shelves for rabbits and shelters for rodents allow them to retreat in case of disturbances (Baumans 1997; Chmiel and Noonan 1996; Stauffacher 1992); and nesting material and deep bedding allow mice to control their temperature and avoid cold stress during resting and sleeping (Gaskill et al. 2009; Gordon 1993, 2004).

Not every item added to the animals' environment benefits their well-being. For example, marbles are used as a stressor in mouse anxiety studies (De Boer and Koolhaas 2003), indicating that some items may be detrimental to well-being. For nonhuman primates, novel objects can increase the risk of disease transmission (Bayne et al. 1993); foraging devices can lead to increased body weight (Brent 1995); shavings can lead to allergies and skin rashes in some individuals; and some objects can result in injury from foreign material in the intestine (Hahn et al. 2000). In some strains of mice, cage dividers and shelters have induced overt aggression in groups of males, resulting in social stress and injury (e.g., Bergmann et al. 1994; Haemisch et al. 1994). Social stress was most likely to occur when resources were monopolized by dominant animals (Bergmann et al. 1994).

Enrichment programs should be reviewed by the IACUC, researchers, and veterinarian on a regular basis to ensure that they are beneficial to animal well-being and consistent with the goals of animal use. They should be updated as needed to ensure that they reflect current knowledge. Personnel responsible for animal care and husbandry should receive training in the behavioral biology of the species they work with to appropriately monitor the effects of enrichment as well as identify the development of adverse or abnormal behaviors.

Like other environmental factors (such as space, light, noise, temperature, and animal care procedures), enrichment affects animal phenotype and may affect the experimental outcome. It should therefore be considered an independent variable and appropriately controlled.

Some scientists have raised concerns that environmental enrichment may compromise experimental standardization by introducing variability, adding not only diversity to the animals' behavioral repertoire but also variation to their responses to experimental treatments (e.g., Bayne 2005; Eskola et al. 1999; Gärtner 1999; Tsai et al. 2003). A systematic study in mice did not find evidence to support this viewpoint (Wolfer et al. 2004), indicating that housing conditions can be enriched without compromising the precision or reproducibility of experimental results. Further research in other species may be needed to confirm this conclusion. However, it has been shown that conditions resulting in higher-stress reactivity increase variation in experimental data (e.g., Macrì et al. 2007). Because adequate environmental enrichment may reduce anxiety and stress reactivity (Chapillon et al. 1999), it may also contribute to higher test sensitivity and reduced animal use (Baumans 1997).

Sheltered or Outdoor Housing

Sheltered or outdoor housing (e.g., barns, corrals, pastures, islands) is a primary housing method for some species and is acceptable in many situations. Animals maintained in outdoor runs, pens, or other large enclosures must have protection from extremes in temperature or other harsh weather conditions and adequate opportunities for retreat (for subordinate animals). These goals can normally be achieved by providing windbreaks, species-appropriate shelters, shaded areas, areas with forced ventilation, heat-radiating structures, and/or means of retreat to conditioned spaces, such as an indoor portion of a run. Shelters should be large enough to accommodate all animals housed in the enclosure, be accessible at all times to all animals, have sufficient ventilation, and be designed to prevent buildup of waste materials and excessive moisture. Houses, dens, boxes, shelves, perches, and other furnishings should be constructed in a manner and made of materials that allow cleaning or replacement in accord with generally accepted husbandry practices.

Floors or ground-level surfaces of outdoor housing facilities may be covered with dirt, absorbent bedding, sand, gravel, grass, or similar material that can be removed or replaced when needed to ensure appropriate sanitation. Excessive buildup of animal waste and stagnant water should be avoided by, for example, using contoured or drained surfaces. Other surfaces should be able to withstand the elements and be easily maintained.

Successful management of outdoor housing relies on stable social groups of compatible animals; sufficient and species-appropriate feeding and resting places; an adequate acclimation period in advance of seasonal changes when animals are first introduced to outdoor housing; training of animals to cooperate with veterinary and investigative personnel (e.g., to enter chutes or cages for restraint or transport); and adequate security via a perimeter fence or other means.

Naturalistic Environments

Areas such as pastures and islands may provide a suitable environment for maintaining or producing animals and for some types of research. Their use results in the loss of some control over nutrition, health care and surveillance, and pedigree management. These limitations should be balanced against the benefits of having the animals live in more natural conditions. Animals should be added to, removed from, and returned to social groups in this setting with appropriate consideration of the effects on the individual animals and on the group. Adequate supplies of food, fresh water, and natural or constructed shelter should be ensured.

Space

General Considerations for All Animals An animal's space needs are complex and consideration of only the animal's body weight or surface area may be inadequate. Important considerations for determining space needs include the age and sex of the animal(s), the number of animals to be cohoused and the duration of the accommodation, the use for which the animals are intended (e.g., production vs. experimentation), and any special needs they may have (e.g., vertical space for arboreal species or thermal gradient for poikilotherms). In many cases, for example, adolescent animals, which usually weigh less than adults but are more active, may require more space relative to body weight (Ikemoto and Panksepp 1992). Group-housed, social animals can share space such that the amount of space required per animal may decrease with increasing group size; thus larger groups may be housed at slightly higher stocking densities than smaller groups or individual animals. Socially housed animals should have sufficient space and structural complexity to allow them to escape aggression or hide from other animals in the pair or group. Breeding animals will require more space, particularly if neonatal animals will be raised together with their mother or as a breeding group until weaning age. Space quality also affects its usability. Enclosures that are complex and environmentally enriched may increase activity and facilitate the expression of species-specific behaviors, thereby increasing space needs. Thus there is no ideal formula for calculating an animal's space needs based only on body size or weight and readers should take the performance indices discussed in this section into consideration when utilizing the species-specific guidelines presented in the following pages.

Consideration of floor area alone may not be sufficient in determining adequate cage size; with some species, cage volume and spatial arrangement may be of greater importance. In this regard, the *Guide* may differ from the US Animal Welfare Regulations (AWRs) or other guidelines. The height of an enclosure can be important to allow for expression of species-specific behaviors and postural adjustments. Cage height should take into account the animal's typical posture and provide adequate clearance for the animal from cage structures, such as feeders and water devices. Some species—for example, nonhuman primates, cats, and arboreal animals—use the vertical dimensions of the cage to a greater extent than the floor. For these animals, the ability to stand or to perch with adequate vertical space to keep their body, including their tail, above the cage floor can improve their well-being (Clarence et al. 2006; MacLean et al. 2009).

Space allocations should be assessed, reviewed, and modified as necessary by the IACUC considering the performance indices (e.g., health, reproduction, growth, behavior, activity, and use of space) and special needs determined by the characteristics of the animal strain or species (e.g., obese, hyperactive, or arboreal animals) and experimental use (e.g., animals in long-term studies may require greater and more complex space). At a minimum, animals must have enough space to express their natural postures and postural adjustments without touching the enclosure walls or ceiling, be able to turn around, and have ready access to food and water. In addition, there must be sufficient space to comfortably rest away from areas soiled by urine and feces. Floor space taken up by food bowls, water containers, litter boxes, and enrichment devices (e.g., novel objects, toys, foraging devices) should not be considered part of the floor space.

The space recommendations presented here are based on professional judgment and experience. They should be considered the minimum for animals housed under conditions commonly found in laboratory animal housing facilities. Adjustments to the amount and arrangement of space recommended in the following tables should be reviewed and approved by the IACUC and should be based on performance indices related to animal well-being and research quality as described in the preceding paragraphs, with due consideration of the AWRs and PHS Policy and other applicable regulations and standards.

It is not within the scope of the *Guide* to discuss the housing requirements of all species used in research. For species not specifically indicated, advice should be sought from the scientific literature and from species-relevant experts.

Laboratory Rodents Table 3.2 lists recommended minimum space for commonly used laboratory rodents housed in groups. If they are housed singly or in small groups or exceed the weights in the table, more space per

TABLE 3.2 Recommended Minimum Space for Commonly Used Laboratory Rodents Housed in Groups*

Animals	Weight, g	Floor Area/Animal,[a] in.^{2} (cm^2)	Height,[b] in. (cm)	Comments
Mice in groups[c]	<10	6 (38.7)	5 (12.7)	Larger animals may require more space to meet the performance standards.
	Up to 15	8 (51.6)	5 (12.7)	
	Up to 25	12 (77.4)	5 (12.7)	
	>25	≥15 (≥96.7)	5 (12.7)	
Female + litter		51 (330) (recommended space for the housing group)	5 (12.7)	Other breeding configurations may require more space and will depend on considerations such as number of adults and litters, and size and age of litters.[d]
Rats in groups[c]	<100	17 (109.6)	7 (17.8)	Larger animals may require more space to meet the performance standards.
	Up to 200	23 (148.35)	7 (17.8)	
	Up to 300	29 (187.05)	7 (17.8)	
	Up to 400	40 (258.0)	7 (17.8)	
	Up to 500	60 (387.0)	7 (17.8)	
	>500	≥70 (≥451.5)	7 (17.8)	
Female + litter		124 (800) (recommended space for the housing group)	7 (17.8)	Other breeding configurations may require more space and will depend on considerations such as number of adults and litters, and size and age of litters.[d]
Hamsters[c]	<60	10 (64.5)	6 (15.2)	Larger animals may require more space to meet the performance standards.
	Up to 80	13 (83.8)	6 (15.2)	
	Up to 100	16 (103.2)	6 (15.2)	
	>100	≥19 (≥122.5)	6 (15.2)	
Guinea pigs[c]	Up to 350	60 (387.0)	7 (17.8)	Larger animals may require more space to meet the performance standards.
	>350	≥101 (≥651.5)	7 (17.8)	

*The interpretation of this table should take into consideration the performance indices described in the text beginning on page 55.

[a]Singly housed animals and small groups may require more than the applicable multiple of the indicated floor space per animal.
[b]From cage floor to cage top.
[c]Consideration should be given to the growth characteristics of the stock or strain as well as the sex of the animal. Weight gain may be sufficiently rapid that it may be preferable to provide greater space in anticipation of the animal's future size. In addition, juvenile rodents are highly active and show increased play behavior.
[d]Other considerations may include culling of litters or separation of litters from the breeding group, as well as other methods of more intensive management of available space to allow for the safety and well-being of the breeding group. Sufficient space should be allocated for mothers with litters to allow the pups to develop to weaning without detrimental effects for the mother or the litter.

animal may be required, while larger groups may be housed at slightly higher densities.

Studies have recently evaluated space needs and the effects of social housing, group size, and density (Andrade and Guimaraes 2003; Bartolomucci et al. 2002, 2003; Georgsson et al. 2001; Gonder and Laber 2007; Perez et al. 1997; A.L. Smith et al. 2004), age (Arakawa 2005; Davidson et al. 2007; Yildiz et al. 2007), and housing conditions (Gordon et al. 1998; Van Loo et al. 2004) for many different species and strains of rodents, and have reported varying effects on behavior (such as aggression) and experimental outcomes (Karolewicz and Paul 2001; Laber et al. 2008; McGlone et al. 2001; Rock et al. 1997; Smith et al. 2005; Van Loo et al. 2001). However, it is difficult to compare these studies due to the study design and experimental variables that have been measured. For example, variables that may affect the animals' response to different cage sizes and housing densities include, but are not limited to, species, strain (and social behavior of the strain), phenotype, age, gender, quality of the space (e.g., vertical access), and structures placed in the cage. These issues remain complex and should be carefully considered when housing rodents.

Other Common Laboratory Animals Tables 3.3 and 3.4 list recommended minimum space for other common laboratory animals and for avian species. These allocations are based, in general, on the needs of pair- or group-housed animals. Space allocations should be reevaluated to provide for enrichment or to accommodate animals that exceed the weights in the tables, and should be based on species characteristics, behavior, compatibility of the animals, number of animals, and goals of the housing situation (Held et al. 1995; Lupo et al. 2000; Raje 1997; Turner et al. 1997). Singly housed animals may require more space per animal than that recommended for group-housed animals, while larger groups may be housed at slightly higher densities. For cats, dogs, and some rabbits, housing enclosures that allow greater freedom of movement and less restricted vertical space are preferred (e.g., kennels, runs, or pens instead of cages). Dogs and cats, especially when housed individually or in smaller enclosures (Bayne 2002), should be allowed to exercise and provided with positive human interaction. Species-specific plans for housing and management should be developed. Such plans should also include strategies for environmental enrichment.

Nonhuman Primates The recommended minimum space for nonhuman primates detailed in Table 3.5 is based on the needs of pair- or group-housed animals. Like all social animals, nonhuman primates should normally have social housing (i.e., in compatible pairs or in larger groups of compatible animals) (Hotchkiss and Paule 2003; NRC 1998a; Weed and Watson 1998;

TABLE 3.3 Recommended Minimum Space for Rabbits, Cats, and Dogs Housed in Pairs or Groups*

Animals	Weight,[a] kg	Floor Area/ Animal,[b] ft^2 (m^2)	Height,[c] in. (cm)	Comments
Rabbits	<2	1.5 (0.14)	16 (40.5)	Larger rabbits may require more cage height to allow animals to sit up.
	Up to 4	3.0 (0.28)	16 (40.5)	
	Up to 5.4	4.0 (0.37)	16 (40.5)	
	>5.4[c]	≥5.0 (≥0.46)	16 (40.5)	
Cats	≤4	3.0 (0.28)	24 (60.8)	Vertical space with perches is preferred and may require additional cage height.
	>4[d]	≥4.0 (≥0.37)	24 (60.8)	
Dogs[e]	<15	8.0 (0.74)	—[f]	Cage height should be sufficient for the animals to comfortably stand erect with their feet on the floor.
	Up to 30	12.0 (1.2)	—[f]	
	>30[d]	≥24.0 (≥2.4)	—[f]	

*The interpretation of this table should take into consideration the performance indices described in the text beginning on page 55.

[a]To convert kilograms to pounds, multiply by 2.2.
[b]Singly housed animals may require more space per animal than recommended for pair- or group-housed animals.
[c]From cage floor to cage top.
[d]Larger animals may require more space to meet performance standards (see text).
[e]These recommendations may require modification according to body conformation of individual animals and breeds. Some dogs, especially those toward the upper limit of each weight range, may require additional space to ensure compliance with the regulations of the Animal Welfare Act. These regulations (USDA 1985) mandate that the height of each cage be sufficient to allow the occupant to stand in a "comfortable position" and that the minimal square feet of floor space be equal to the "mathematical square of the sum of the length of the dog in inches (measured from the tip of its nose to the base of its tail) plus 6 inches; then divide the product by 144."
[f]Enclosures that allow greater freedom of movement and unrestricted height (i.e., pens, runs, or kennels) are preferable.

Wolfensohn 2004). Group composition is critical and numerous species-specific factors such as age, behavioral repertoire, sex, natural social organization, breeding requirements, and health status should be taken into consideration when forming a group. In addition, due to conformational differences of animals within groups, more space or height may be required to meet the animals' physical and behavioral needs. Therefore, determination of the appropriate cage size is not based on body weight alone, and professional judgment is paramount in making such determinations (Kaufman et al. 2004; Williams et al. 2000).

TABLE 3.4 Recommended Minimum Space for Avian Species Housed in Pairs or Groups*

Animals	Weight,[a] kg	Floor area/animal,[b] ft^2 (m^2)	Height
Pigeons	—	0.8 (0.07)	Cage height should be sufficient for the animals to comfortably stand erect with their feet on the floor.
Quail	—	0.25 (0.023)	
Chickens	<0.25	0.25 (0.023)	
	Up to 0.5	0.50 (0.046)	
	Up to 1.5	1.00 (0.093)	
	Up to 3.0	2.00 (0.186)	
	>3.0[c]	≥3.00 (≥0.279)	

*The interpretation of this table should take into consideration the performance indices described in the text beginning on page 55.

[a]To convert kilograms to pounds, multiply by 2.2.
[b]Singly housed birds may require more space per animal than recommended for pair- or group-housed birds.
[c]Larger animals may require more space to meet performance standards (see text).

If it is necessary to house animals singly—for example, when justified for experimental purposes, for provision of veterinary care, or for incompatible animals—this arrangement should be for the shortest duration possible. If single animals are housed in small enclosures, an opportunity for periodic release into larger enclosures with additional enrichment items should be considered, particularly for animals housed singly for extended periods of time. Singly housed animals may require more space per animal than recommended for pair- or group-housed animals, while larger groups may be housed at slightly higher densities. Because of the many physical and behavioral characteristics of nonhuman primate species and the many factors to consider when using these animals in a biomedical research setting, species-specific plans for housing and management should be developed. Such plans should include strategies for environmental and psychological enrichment.

Agricultural Animals Table 3.6 lists recommended minimum space for agricultural animals commonly used in a laboratory setting. As social animals, they should be housed in compatible pairs or larger groups of compatible animals. When animals exceed the weights in the table, more space is required. For larger animals (particularly swine) it is important that the configuration of the space allow the animals to turn around and move freely (Becker et al. 1989; Bracke et al. 2002). Food troughs and water devices should be provided in sufficient numbers to allow ready access for all animals. Singly housed animals may require more space than recommended in

TABLE 3.5 Recommended Minimum Space for Nonhuman Primates Housed in Pairs or Groups*

Animals	Weight,[a] kg	Floor area/animal,[b] ft^2 (m^2)	Height,[c] in. (cm)	Comments
Monkeys[d] (including baboons)				Cage height should be sufficient for the animals to comfortably stand erect with their feet on the floor. Baboons, patas monkeys, and other longer-legged species may require more height than other monkeys, as might long-tailed animals and animals with prehensile tails. Overall cage volume and linear perch space should be considerations for many neotropical and arboreal species. For brachiating species cage height should be such that an animal can, when fully extended, swing from the cage ceiling without having its feet touch the floor. Cage design should enhance brachiating movement.
Group 1	Up to 1.5	2.1 (0.20)	30 (76.2)	
Group 2	Up to 3	3.0 (0.28)	30 (76.2)	
Group 3	Up to 10	4.3 (0.4)	30 (76.2)	
Group 4	Up to 15	6.0 (0.56)	32 (81.3)	
Group 5	Up to 20	8.0 (0.74)	36 (91.4)	
Group 6	Up to 25	10 (0.93)	46 (116.8)	
Group 7	Up to 30	15 (1.40)	46 (116.8)	
Group 8	>30[e]	≥25 (≥2.32)	60 (152.4)	
Chimpanzees (*Pan*)				For other apes and large brachiating species cage height should be such that an animal can, when fully extended, swing from the cage ceiling without having its feet touch the floor. Cage design should enhance brachiating movement.
Juveniles	Up to 10	15 (1.4)	60 (152.4)	
Adults[f]	>10	≥25 (≥2.32)	84 (213.4)	

*The interpretation of this table should take into consideration the performance indices described in the text beginning on page 55.

[a]To convert kilograms to pounds, multiply by 2.2.
[b]Singly housed primates may require more space than the amount allocated per animal when group housed.
[c]From cage floor to cage top.
[d]Callitrichidae, Cebidae, Cercopithecidae, and Papio.
[e]Larger animals may require more space to meet performance standards (see text).
[f]Apes weighing over 50 kg are more effectively housed in permanent housing of masonry, concrete, and wire-panel structure than in conventional caging.

TABLE 3.6 Recommended Minimum Space for Agricultural Animals*

Animals/Enclosure	Weight,[a] kg	Floor Area/Animal,[b] ft² (m²)
Sheep and Goats		
1	<25	10.0 (0.9)
	Up to 50	15.0 (1.35)
	>50[c]	≥20.0 (≥1.8)
2-5	<25	8.5 (0.76)
	Up to 50	12.5 (1.12)
	>50[c]	≥17.0 (≥1.53)
>5	<25	7.5 (0.67)
	Up to 50	11.3 (1.02)
	>50[c]	≥15.0 (≥1.35)
Swine		
1	<15	8.0 (0.72)
	Up to 25	12.0 (1.08)
	Up to 50	15.0 (1.35)
	Up to 100	24.0 (2.16)
	Up to 200	48.0 (4.32)
	>200[c]	≥60.0 (≥5.4)
2-5	<25	6.0 (0.54)
	Up to 50	10.0 (0.9)
	Up to 100	20.0 (1.8)
	Up to 200	40.0 (3.6)
	>200[c]	≥52.0 (≥4.68)
>5	<25	6.0 (0.54)
	Up to 50	9.0 (0.81)
	Up to 100	18.0 (1.62)
	Up to 200	36.0 (3.24)
	>200[c]	≥48.0 (≥4.32)
Cattle		
1	<75	24.0 (2.16)
	Up to 200	48.0 (4.32)
	Up to 350	72.0 (6.48)
	Up to 500	96.0 (8.64)
	Up to 650	124.0 (11.16)
	>650[c]	≥144.0 (≥12.96)
2-5	<75	20.0 (1.8)
	Up to 200	40.0 (3.6)
	Up to 350	60.0 (5.4)
	Up to 500	80.0 (7.2)
	Up to 650	105.0 (9.45)
	>650[c]	≥120.0 (≥10.8)

>5	<75	18.0 (1.62)
	Up to 200	36.0 (3.24)
	Up to 350	54.0 (4.86)
	Up to 500	72.0 (6.48)
	Up to 650	93.0 (8.37)
	>650[c]	≥108.0 (≥9.72)
Horses	—	144.0 (12.96)
Ponies		
1-4	—	72.0 (6.48)
>4/Pen	≤200	60.0 (5.4)
	>200[c]	≥72.0 (≥6.48)

*The interpretation of this table should take into consideration the performance indices described in the text beginning on page 55.

[a]To convert kilograms to pounds, multiply by 2.2.
[b]Floor area configuration should be such that animals can turn around and move freely without touching food or water troughs, have ready access to food and water, and have sufficient space to comfortably rest away from areas soiled by urine and feces (see text).
[c]Larger animals may require more space to meet performance standards including sufficient space to turn around and move freely (see text).

the table to enable them to turn around and move freely without touching food or water troughs, have ready access to food and water, and have sufficient space to comfortably rest away from areas soiled by urine and feces.

Terrestrial Management

Behavioral and Social Management

Activity Animal Activity typically implies motor activity but also includes cognitive activity and social interaction. Animals' natural behavior and activity profile should be considered during evaluation of suitable housing or behavioral assessment.

Animals maintained in a laboratory environment are generally restricted in their activities compared to free-ranging animals. Forced activity for reasons other than attempts to meet therapeutic or approved protocol objectives should be avoided. High levels of repetitive, unvarying behavior (stereotypies, compulsive behaviors) may reflect disruptions of normal behavioral control mechanisms due to housing conditions or management practices (Garner 2005; NRC 1998a).

Dogs, cats, rabbits, and many other animals benefit from positive human interaction (Augustsson et al. 2002; Bayne et al. 1993; McCune 1997; Poole 1998; Rennie and Buchanan-Smith 2006; Rollin 1990). Dogs can be given additional opportunities for activity by being walked on a leash, having access to a run, or being moved into areas for social contact, play, or exploration (Wolff and Rupert 1991). Loafing areas, exercise lots, and pastures are suitable for large farm animals, such as sheep, horses, and cattle.

Social Environment Appropriate social interactions among members of the same species (conspecifics) are essential to normal development and well-being (Bayne et al. 1995; Hall 1998; Novak et al. 2006). When selecting a suitable social environment, attention should be given to whether the animals are naturally territorial or communal and whether they should be housed singly, in pairs, or in groups. An understanding of species-typical natural social behavior (e.g., natural social composition, population density, ability to disperse, familiarity, and social ranking) is key to successful social housing.

Not all members of a social species are necessarily socially compatible. Social housing of incompatible animals can induce chronic stress, injury, and even death. In some species, social incompatibility may be sex biased; for example, male mice are generally more prone to aggression than female mice, and female hamsters are generally more aggressive than male hamsters. Risks of social incompatibility are greatly reduced if the animals to be grouped are raised together from a young age, if group composition remains stable, and if the design of the animals' enclosure and their environmental enrichment facilitate the avoidance of social conflicts. Social stability should be carefully monitored; in cases of severe or prolonged aggression, incompatible individuals need to be separated.

For some species, developing a stable social hierarchy will entail antagonistic interactions between pair or group members, particularly for animals introduced as adults. Animals may have to be introduced to each other over a period of time and should be monitored closely during this introductory period and thereafter to ensure compatibility.

Single housing of social species should be the exception and justified based on experimental requirements or veterinary-related concerns about animal well-being. In these cases, it should be limited to the minimum period necessary, and where possible, visual, auditory, olfactory, and tactile contact with compatible conspecifics should be provided. In the absence of other animals, enrichment should be offered such as positive interaction with the animal care staff and additional enrichment items or addition of a companion animal in the room or housing area. The need for single housing should be reviewed on a regular basis by the IACUC and veterinarian.

Procedural Habituation and Training of Animals Habituating animals to routine husbandry or experimental procedures should be encouraged whenever possible as it may assist the animal to better cope with a captive environment by reducing stress associated with novel procedures or people.

The type and duration of habituation needed will be determined by the complexity of the procedure. In most cases, principles of operant conditioning may be employed during training sessions, using progressive behavioral shaping, to induce voluntary cooperation with procedures (Bloomsmith et al. 1998; Laule et al. 2003; NRC 2006a; Reinhardt 1997).

Husbandry

Food Animals should be fed palatable, uncontaminated diets that meet their nutritional and behavioral needs at least daily, or according to their particular requirements, unless the protocol in which they are being used requires otherwise. Subcommittees of the National Research Council Committee on Animal Nutrition have prepared comprehensive reports of the nutrient requirements of laboratory animals (NRC 1977, 1982, 1993, 1994, 1995a, 1998b, 2000, 2001, 2003a, 2006b,c, 2007); these publications consider issues of quality assurance, freedom from chemical or microbial contaminants and natural toxicants in feedstuffs, bioavailability of nutrients in feeds, and palatability.

There are several types of diets classified by the degree of refinement of their ingredients. *Natural-ingredient diets* are formulated with agricultural products and byproducts and are commercially available for all species commonly used in the laboratory. Although not a significant factor in most instances, the nutrient composition of ingredients varies, and natural ingredients may contain low levels of naturally occurring or artificial contaminants (Ames et al. 1993; Knapka 1983; Newberne 1975; NRC 1996; Thigpen et al. 1999, 2004). Contaminants such as pesticide residues, heavy metals, toxins, carcinogens, and phytoestrogens may be at levels that induce few or no health sequelae yet may have subtle effects on experimental results (Thigpen et al. 2004). *Certified diets* that have been assayed for contaminants are commercially available for use in select studies, such as preclinical toxicology, conducted in compliance with FDA Good Laboratory Practice standards (CFR 2009). *Purified diets* are refined such that each ingredient contains a single nutrient or nutrient class; they have less nutrient concentration variability and the potential for chemical contamination is lower. *Chemically defined diets* contain the most elemental ingredients available, such as individual amino acids and specific sugars (NRC 1996). The latter two types of diet are more likely to be used for specific types of studies in rodents but are not commonly used because of cost, lower palatability, and a reduced shelf life.

Animal colony managers should be judicious when purchasing, transporting, storing, and handling food to minimize the introduction of diseases, parasites, potential disease vectors (e.g., insects and other vermin), and chemical contaminants in animal colonies. Purchasers are encouraged to consider manufacturers' and suppliers' procedures and practices (e.g., storage, vermin control, and handling) for protecting and ensuring diet quality.

Institutions should urge feed vendors to periodically provide data from laboratory-based feed analyses for critical nutrients. The user should know the date of manufacture and other factors that affect the food's shelf life. Stale food or food transported and stored inappropriately can become deficient in nutrients. Upon receipt, bags of feed should be examined to ensure that they are intact and unstained to help ensure that their contents have not been potentially exposed to vermin, penetrated by liquids, or contaminated. Careful attention should be paid to quantities received in each shipment, and stock should be rotated so that the oldest food is used first.

Areas in which diets and diet ingredients are processed or stored should be kept clean and enclosed to prevent the entry of pests. Food stocks should be stored off the floor on pallets, racks, or carts in a manner that facilitates sanitation. Opened bags of food should be stored in vermin-proof containers to minimize contamination and to avoid the potential spread of pathogens. Exposure to elevated storage room temperatures, extremes in relative humidity, unsanitary conditions, and insects and other vermin hastens food deterioration. Storage of natural-ingredient diets at less than 21°C (70°F) and below 50% relative humidity is recommended. Precautions should be taken if perishable items—such as meats, fruits, and vegetables and some specialty diets (e.g., select medicated or high-fat diets)—are fed, because storage conditions may lead to variation in food quality.

Most natural-ingredient, dry laboratory animal diets stored properly can be used up to 6 months after manufacture. Nonstabilized vitamin C in manufactured feeds generally has a shelf life of only 3 months, but commonly used stabilized forms can extend the shelf life of feed. Refrigeration preserves nutritional quality and lengthens shelf life, but food storage time should be reduced to the lowest practical period and the manufacturers' recommendations considered. Purified and chemically defined diets are often less stable than natural-ingredient diets and their shelf life is usually less than 6 months (Fullerton et al. 1982); they should be stored at 4°C (39°F) or lower.

Irradiated and fortified autoclavable diets are commercially available and are commonly used for axenic and microbiologically defined rodents, and immunodeficient animals (NRC 1996). The use of commercially fortified autoclavable diets ensures that labile vitamin content is not compromised by steam and/or heat (Caulfield et al. 2008; NRC 1996). But consideration should be given to the impact of autoclaving on pellets as it may affect their hardness and thus palatability and also lead to chemical alteration of ingredients (Thigpen et al. 2004; Twaddle et al. 2004). The date of sterilization should be recorded and the diet used quickly.

Feeders should be designed and placed to allow easy access to food and to minimize contamination with urine and feces, and maintained in good condition. When animals are housed in groups, there should be enough space and enough feeding points to minimize competition for food and ensure access to food for all animals, especially if feed is restricted as part of the protocol or management routine. Food storage containers should not be transferred between areas that pose different risks of contamination without appropriate treatment, and they should be cleaned and sanitized regularly.

Management of caloric intake is an accepted practice for long-term housing of some species, such as some rodents, rabbits, and nonhuman primates, and as an adjunct to some clinical, experimental, and surgical procedures (for more discussion of food and fluid regulation as an experimental tool see Chapter 2 and NRC 2003a). Benefits of moderate caloric restriction in some species may include increased longevity and reproduction, and decreased obesity, cancer rates, and neurogenerative disorders (Ames et al. 1993; Colman et al. 2009; Keenan et al. 1994, 1996; Lawler et al. 2008; Weindruch and Walford 1988).

Under standard housing conditions, changes in biologic needs commensurate with aging should be taken into consideration. For example, there is good evidence that mice and rats with continuous access to food can become obese, with attendant metabolic and cardiovascular changes such as insulin resistance and higher blood pressure (Martin et al. 2010). These and other changes along with a more sedentary lifestyle and lack of exercise increase the risk of premature death (ibid.). Caloric management, which may affect physiologic adaptations and alter metabolic responses in a species-specific manner (Leveille and Hanson 1966), can be achieved by reducing food intake or by stimulating exercise.

In some species (e.g., nonhuman primates) and on some occasions, varying nutritionally balanced diets and providing "treats," including fresh fruit and vegetables, can be appropriate and improve well-being. Scattering food in the bedding or presenting part of the diet in ways that require the animals to work for it (e.g., puzzle feeders for nonhuman primates) gives the animals the opportunity to forage, which, in nature, normally accounts for a large proportion of their daily activity. A diet should be nutritionally balanced; it is well documented that many animals offered a choice of unbalanced or balanced foods do not select a balanced diet and become malnourished or obese through selection of high-energy, low-protein foods (Moore 1987). Abrupt changes in diet, which can be difficult to avoid at weaning, should be minimized because they can lead to digestive and metabolic disturbances; these changes occur in omnivores and carnivores, but herbivores (Eadie and Mann 1970) are especially sensitive.

Water Animals should have access to potable, uncontaminated drinking water according to their particular requirements. Water quality and the definition of potable water can vary with locality (Homberger et al. 1993). Periodic monitoring for pH, hardness, and microbial or chemical contamination may be necessary to ensure that water quality is acceptable, particularly for use in studies in which normal components of water in a given locality can influence the results. Water can be treated or purified to minimize or eliminate contamination when protocols require highly purified water. The selection of water treatments should be carefully considered because many forms of water treatment have the potential to cause physiologic alterations, reduction in water consumption, changes in microflora, or effects on experimental results (Fidler 1977; Hall et al. 1980; Hermann et al. 1982; Homberger et al. 1993; NRC 1996).

Watering devices, such as drinking tubes and automated water delivery systems, should be checked frequently to ensure appropriate maintenance, cleanliness, and operation. Animals sometimes have to be trained to use automated watering devices and should be observed regularly until regular usage has been established to prevent dehydration. It is better to replace water bottles than to refill them, because of the potential for microbiologic cross contamination; if bottles are refilled, care should be taken to return each bottle to the cage from which it was removed. Automated watering distribution systems should be flushed or disinfected regularly. Animals housed in outdoor facilities may have access to water in addition to that provided in watering devices, such as that available in streams or in puddles after a heavy rainfall. Care should be taken to ensure that such accessory sources of water do not constitute a hazard, but their availability need not routinely be prevented. In cold weather, steps should be taken to prevent freezing of outdoor water sources.

Bedding and Nesting Materials Animal bedding and nesting materials are controllable environmental factors that can influence experimental data and improve animal well-being in most terrestrial species. Bedding is used to absorb moisture, minimize the growth of microorganisms, and dilute and limit animals' contact with excreta, and specific bedding materials have been shown to reduce the accumulation of intracage ammonia (Perkins and Lipman 1995; E. Smith et al. 2004). Various materials are used as both contact and noncontact bedding; the desirable characteristics and methods of evaluating bedding have been described (Gibson et al. 1987; Jones 1977; Kraft 1980; Thigpen et al. 1989; Weichbrod et al. 1986). The veterinarian or facility manager, in consultation with investigators, should select the most appropriate bedding and nesting materials. A number of species, most notably rodents, exhibit a clear preference for specific materials (Blom et al. 1996; Manser et al. 1997, 1998; Ras et al. 2002), and mice provided with appropriate nesting material build better nests (Hess et al. 2008). Bedding that enables burrowing is encouraged for some species, such as mice and hamsters.

No type of bedding is ideal for all species under all management and experimental conditions. For example, in nude or hairless mice that lack eyelashes, some forms of paper bedding with fines (i.e., very small particles found in certain types of bedding) can result in periorbital abscesses (White et al. 2008), while cotton nestlets may lead to conjunctivitis (Bazille et al. 2001). Bedding can also influence mucosal immunity (Sanford et al. 2002) and endocytosis (Buddaraju and Van Dyke 2003).

Softwood beddings have been used, but the use of untreated softwood shavings and chips is contraindicated for some protocols because they can affect metabolism (Vesell 1967; Vesell et al. 1973, 1976). Cedar shavings are not recommended because they emit aromatic hydrocarbons that induce hepatic microsomal enzymes and cytotoxicity (Torronen et al. 1989; Weichbrod et al. 1986, 1988) and have been reported to increase the incidence of cancer (Jacobs and Dieter 1978; Vlahakis 1977). Prior treatment with high heat (kiln drying or autoclaving) may, depending on the material and the concentration of aromatic hydrocarbon constituents, reduce the concentration of volatile organic compounds, but the amounts remaining may be sufficient to affect specific protocols (Cunliffe-Beamer et al. 1981; Nevalainen and Vartiainen 1996).

The purchase of bedding products should take into consideration vendors' manufacturing, monitoring, and storage methods. Bedding may be contaminated with toxins and other substances, bacteria, fungi, and vermin. It should be transported and stored off the floor on pallets, racks, or carts in a fashion consistent with maintenance of quality and avoidance of contamination. Bags should be stored sufficiently away from walls to facilitate cleaning. During autoclaving, bedding can absorb moisture and as a result lose absorbency and support the growth of microorganisms. Therefore, appropriate drying times and storage conditions should be used or, alternatively, gamma-irradiated materials if sterile bedding is indicated.

Bedding should be used in amounts sufficient to keep animals dry between cage changes, and, in the case of small laboratory animals, it should be kept from coming into contact with sipper tubes as such contact could cause leakage of water into the cage.

Sanitation *Sanitation*—the maintenance of environmental conditions conducive to health and well-being—involves bedding change (as appropriate), cleaning, and disinfection. *Cleaning* removes excessive amounts of excrement, dirt, and debris, and *disinfection* reduces or eliminates unacceptable concentrations of microorganisms. The goal of any sanitation program is to maintain sufficiently clean and dry bedding, adequate air quality, and clean cage surfaces and accessories.

The frequency and intensity of cleaning and disinfection should depend on what is necessary to provide a healthy environment for an animal. Methods and frequencies of sanitation will vary with many factors, including the normal physiologic and behavioral characteristics of the animals; the type, physical characteristics, and size of the enclosure; the type, number, size, age, and reproductive status of the animals; the use and type of bedding materials; temperature and relative humidity; the nature of the materials that create the need for sanitation; and the rate of soiling of the surfaces of the enclosure. Some housing systems or experimental protocols may require specific husbandry techniques, such as aseptic handling or modification in the frequency of bedding change.

Agents designed to mask animal odors should not be used in animal housing facilities. They cannot substitute for good sanitation practices or for the provision of adequate ventilation, and they expose animals to volatile compounds that might alter basic physiologic and metabolic processes.

Bedding/Substrate Change Soiled bedding should be removed and replaced with fresh materials as often as necessary to keep the animals clean and dry and to keep pollutants, such as ammonia, at a concentration below levels irritating to mucous membranes. The frequency of bedding change depends on multiple factors, such as species, number, and size of the animals in the primary enclosure; type and size of the enclosure; macro- and microenvironmental temperature, relative humidity, and direct ventilation of the enclosure; urinary and fecal output and the appearance and wetness of bedding; and experimental conditions, such as those of surgery or debilitation, that might limit an animal's movement or access to clean bedding. There is no absolute minimal frequency of bedding changes; the choice is a matter of professional judgment and consultation between the investigator and animal care personnel. It typically varies from daily to weekly. In some instances frequent bedding changes are contraindicated; examples include portions of the pre- or postpartum period, research objectives that will be affected, and species in which scent marking is critical and successful reproduction is pheromone dependent.

Cleaning and Disinfection of the Microenvironment The frequency of sanitation of cages, cage racks, and associated equipment (e.g., feeders and watering devices) is governed to some extent by the types of caging and husbandry practices used, including the use of regularly changed contact or noncontact bedding, regular flushing of suspended catch pans, and the use of wire-bottom or perforated-bottom cages. In general, enclosures and accessories, such as tops, should be sanitized at least once every 2 weeks. Solid-bottom caging, bottles, and sipper tubes usually require sanitation at least once a week. Some types of cages and housing systems may require less frequent cleaning or disinfection; such housing may include large cages with very low animal density and frequent bedding changes, cages containing animals in gnotobiotic conditions with frequent bedding changes, individually ventilated cages, and cages used for special situations. Other circumstances, such as filter-topped cages without forced-air ventilation, animals that urinate excessively (e.g., diabetic or renal patients), or densely populated enclosures, may require more frequent sanitation.

The increased use of individually ventilated cages (IVCs) for rodents has led to investigations of the maintenance of a suitable microenvironment with extended cage sanitation intervals and/or increased housing densities (Carissimi et al. 2000; Reeb-Whitaker et al. 2001; Schondelmeyer et al. 2006). By design, ventilated caging systems provide direct continuous exchange of air, compared to static caging systems that depend on passive ventilation from the macroenvironment. As noted above, decreased sanitation frequency may be justified if the microenvironment in the cages, under the conditions of use (e.g., cage type and manufacturer, bedding, species, strain, age, sex, density, and experimental considerations), is not compromised (Reeb et al. 1998). Verification of microenvironmental conditions may include measurement of pollutants such as ammonia and CO_2, microbiologic load, observation of the animals' behavior and appearance, and the condition of bedding and cage surfaces.

Primary enclosures can be disinfected with chemicals, hot water, or a combination of both.[2] Washing times and conditions and postwashing processing procedures (e.g., sterilization) should be sufficient to reduce levels or eliminate vegetative forms of opportunistic and pathogenic bacteria, adventitious viruses, and other organisms that are presumed to be controllable by the sanitation program. Disinfection from the use of hot water alone is the result of the combined effect of the temperature and the length of time that a given temperature (cumulative heat factor) is applied to the surface of the item. The same cumulative heat factor can be obtained by exposing organisms either to very high temperatures for short periods or to lower temperatures for longer periods (Wardrip et al. 1994, 2000). Effective disinfection can be achieved with wash and rinse water at 143-180°F or more. The traditional 82.2°C (180°F) temperature requirement for rinse water refers to the water in the tank or in the sprayer manifold. Detergents and chemical disinfectants enhance the effectiveness of hot water but should be thoroughly rinsed from surfaces before reuse of the equipment. Their use may be contraindicated for some aquatic species, as residue may be highly deleterious. Mechanical washers (e.g., cage and rack, tunnel, and bottle washers) are recommended for cleaning quantities of caging and movable equipment.

Sanitation of cages and equipment by hand with hot water and detergents or disinfectants can also be effective but requires considerable attention to detail. It is particularly important to ensure that surfaces are rinsed free of residual chemicals and that personnel have appropriate equipment to protect themselves from exposure to hot water or chemical agents used in the process.

Water bottles, sipper tubes, stoppers, feeders, and other small pieces of equipment should be washed with detergents and/or hot water and, where

[2]Rabbits and some rodents, such as guinea pigs and hamsters, produce urine with high concentrations of proteins and minerals. These compounds often adhere to cage surfaces and necessitate treatment with acid solutions before and/or during washing.

appropriate, chemical agents to destroy microorganisms. Cleaning with ultrasound may be a useful method for small pieces of equipment.

If automated watering systems are used, some mechanism to ensure that microorganisms and debris do not build up in the watering devices is recommended (Meier et al. 2008); the mechanism can be periodic flushing with large volumes of water or appropriate chemical agents followed by a thorough rinsing. Constant recirculation loops that use properly maintained filters, ultraviolet lights, or other devices to disinfect recirculated water are also effective. Attention should be given to the routine sanitation of automatic water delivery valves (i.e., lixits) during primary enclosure cleaning.

Conventional methods of cleaning and disinfection are adequate for most animal care equipment. However, it may be necessary to also sterilize caging and associated equipment to ensure that pathogenic or opportunistic microorganisms are not introduced into specific-pathogen-free or immunocompromised animals, or that experimental biologic hazards are destroyed before cleaning. Sterilizers should be regularly evaluated and monitored to ensure their safety and effectiveness.

For pens or runs, frequent flushing with water and periodic use of detergents or disinfectants are usually appropriate to maintain sufficiently clean surfaces. If animal waste is to be removed by flushing, this will need to be done at least once a day. During flushing, animals should be kept dry. The timing of pen or run cleaning should take into account the normal behavioral and physiologic processes of the animals; for example, the gastrocolic reflex in meal-fed animals results in defecation shortly after food consumption.

Cleaning and Disinfection of the Macroenvironment All components of the animal facility, including animal rooms and support spaces (e.g., storage areas, cage-washing facilities, corridors, and procedure rooms) should be regularly cleaned and disinfected as appropriate to the circumstances and at a frequency based on the use of the area and the nature of likely contamination. Vaporized hydrogen peroxide or chlorine dioxide are effective compounds for room decontamination, particularly following completion of studies with highly infectious agents (Krause et al. 2001) or contamination with adventitious microbial agents.

Cleaning implements should be made of materials that resist corrosion and withstand regular sanitation. They should be assigned to specific areas and should not be transported between areas with different risks of contamination without prior disinfection. Worn items should be replaced regularly. The implements should be stored in a neat, organized fashion that facilitates drying and minimizes contamination or harborage of vermin.

Assessing the Effectiveness of Sanitation Monitoring of sanitation practices should fit the process and materials being cleaned and may include visual inspection and microbiologic and water temperature monitoring (Compton et al. 2004a,b; Ednie et al. 1998; Parker et al. 2003). The intensity of animal odors, particularly that of ammonia, should not be used as the sole means of assessing the effectiveness of the sanitation program. A decision to alter the frequency of cage bedding changes or cage washing should be based on such factors as ammonia concentration, bedding condition, appearance of the cage and animals, and the number and size of animals housed in the cage.

Mechanical washer function should be evaluated regularly and include examination of mechanical components such as spray arms and moving headers as well as spray nozzles to ensure that they are functioning appropriately. If sanitation is temperature dependent, the use of temperature-sensing devices (e.g., thermometers, probes, or temperature-sensitive indicator strips) is recommended to ensure that the equipment being sanitized is exposed to the desired conditions.

Whether the sanitation process is automated or manual, regular evaluation of sanitation effectiveness is recommended. This can be performed by evaluating processed materials by microbiologic culture or the use of organic material detection systems (e.g., adenosine triphosphate [ATP] bioluminescence) and/or by confirming the removal of artificial soil applied to equipment surfaces before washing.

Waste Disposal Conventional, biologic, and hazardous waste should be removed and disposed of regularly and safely (Hill 1999). There are several options for effective waste disposal. Contracts with licensed commercial waste disposal firms usually provide some assurance of regulatory compliance and safety. On-site incineration should comply with all federal, state, and local regulations (Nadelkov 1996).

Adequate numbers of properly labeled waste receptacles should be strategically placed throughout the facility. Waste containers should be leakproof and equipped with tight-fitting lids. It is good practice to use disposable liners and to wash containers and implements regularly. There should be a dedicated waste storage area that can be kept free of insects and other vermin. If cold storage is used to hold material before disposal, a properly labeled, dedicated refrigerator, freezer, or cold room should be used that is readily sanitized.

Hazardous wastes must be rendered safe by sterilization, containment, or other appropriate means before their removal from the facility (DHHS 2009 or most recent edition; NRC 1989, 1995b). Radioactive wastes should be kept in properly labeled containers and their disposal closely coordinated with radiation safety specialists in accord with federal and state regulations; the federal government and most states and municipalities have regulations controlling disposal of hazardous wastes. Compliance with regulations concerning hazardous-agent use (see Chapter 2) and disposal is an institutional responsibility.

Infectious animal carcasses can be incinerated on site or collected by a licensed contractor. Use of chemical digesters (alkaline hydrolysis treatment) may be considered in some situations (Kaye et al. 1998; Murphy et al. 2009). Procedures for on-site packaging, labeling, transportation, and storage of these wastes should be integrated into occupational health and safety policies (Richmond et al. 2003).

Hazardous wastes that are toxic, carcinogenic, flammable, corrosive, reactive, or otherwise unstable should be placed in properly labeled containers and disposed of as recommended by occupational health and safety specialists. In some circumstances, these wastes can be consolidated or blended. Sharps and glass should be disposed of in a manner that will prevent injury to waste handlers.

Pest Control Programs designed to prevent, control, or eliminate the presence of or infestation by pests are essential in an animal environment. A regularly scheduled and documented program of control and monitoring should be implemented. The ideal program prevents the entry of vermin and eliminates their harborage in the facility (Anadon et al. 2009; Easterbrook et al. 2008). For animals in outdoor facilities, consideration should be given to eliminating or minimizing the potential risk associated with pests and predators.

Pesticides can induce toxic effects on research animals and interfere with experimental procedures (Gunasekara et al. 2008). They should be used in animal areas only when necessary and investigators whose animals may be exposed to them should be consulted beforehand. Use of pesticides should be recorded and coordinated with the animal care management staff and be in compliance with federal, state, or local regulations. Whenever possible, nontoxic means of pest control, such as insect growth regulators (Donahue et al. 1989; Garg and Donahue 1989; King and Bennett 1989; Verma 2002) and nontoxic substances (e.g., amorphous silica gel), should be used. If traps are used, methods should be humane; traps that catch pests alive require frequent observation and humane euthanasia after capture (Mason and Littin 2003; Meerburg et al. 2008).

Emergency, Weekend, and Holiday Care Animals should be cared for by qualified personnel every day, including weekends and holidays, both to safeguard their well-being and to satisfy research requirements. Emergency veterinary care must be available after work hours, on weekends, and on holidays.

In the event of an emergency, institutional security personnel and fire or police officials should be able to reach people responsible for the animals. Notification can be enhanced by prominently posting emergency procedures, names, or telephone numbers in animal facilities or by placing them in the security department or telephone center. Emergency procedures for handling special facilities or operations should be prominently posted and personnel trained in emergency procedures for these areas. A disaster plan that takes into account both personnel and animals should be prepared as part of the overall safety plan for the animal facility. The colony manager or veterinarian responsible for the animals should be a member of the appropriate safety committee at the institution, an "official responder" in the institution, and a participant in the response to a disaster (Vogelweid 1998).

Population Management

Identification Animal records are useful and variable, ranging from limited information on identification cards to detailed computerized records for individual animals (Field et al. 2007). Means of animal identification include room, rack, pen, stall, and cage cards with written, bar-coded, or radio frequency identification (RFID) information. Identification cards should include the source of the animal, the strain or stock, names and contact information for the responsible investigator(s), pertinent dates (e.g., arrival date, birth date, etc.), and protocol number when applicable. Genotype information, when applicable, should also be included, and consistent, unambiguous abbreviations should be used when the full genotype nomenclature (see below) is too lengthy.

In addition, the animals may wear collars, bands, plates, or tabs or be marked by colored stains, ear notches/punches and tags, tattoos, subcutaneous transponders, and freeze brands. As a method of identification of small rodents, toe-clipping should be used only when no other individual identification method is feasible. It may be the preferred method for neonatal mice up to 7 days of age as it appears to have few adverse effects on behavior and well-being at this age (Castelhano-Carlos et al. 2010; Schaefer et al. 2010), especially if toe clipping and genotyping can be combined. Under all circumstances aseptic practices should be followed. Use of anesthesia or analgesia should be commensurate with the age of the animals (Hankenson et al. 2008).

Recordkeeping Records containing basic descriptive information are essential for management of colonies of large long-lived animals and should be maintained for each animal (Dyke 1993; Field et al. 2007; NRC 1979a). These records often include species, animal identifier, sire and/or dam identifier, sex, birth or acquisition date, source, exit date, and final disposition. Such animal records are essential for genetic management and historical assessments of colonies. Records of rearing and housing histories, mating histories, and behavioral profiles are useful for the management of many species, especially nonhuman primates (NRC 1979a). Relevant recorded information should be provided when animals are transferred between institutions.

Medical records for individual animals can also be valuable, especially for dogs, cats, nonhuman primates, and agricultural animals (Suckow and Doerning 2007). They should include pertinent clinical and diagnostic information, date of inoculations, history of surgical procedures and postoperative care, information on experimental use, and necropsy findings where applicable.

Basic demographic information and clinical histories enhance the value of individual animals for both breeding and research and should be readily accessible to investigators, veterinary staff, and animal care staff.

Breeding, Genetics, and Nomenclature Genetic characteristics are important with regard to the selection and management of animals for use in breeding colonies and in biomedical research (see Appendix A). Pedigree information allows appropriate selection of breeding pairs and of experimental animals that are unrelated or of known relatedness.

Outbred animals are widely used in biomedical research. Founding populations should be large enough to ensure the long-term genetic heterogeneity of breeding colonies. To facilitate direct comparison of research data derived from outbred animals, genetic management techniques should be used to maintain genetic variability and equalize founder representations (Hartl 2000; Lacy 1989; Poiley 1960; Williams-Blangero 1991). Genetic variability can be monitored with computer simulations, biochemical markers, DNA markers and sequencing, immunologic markers, or quantitative genetic analyses of physiologic variables (MacCluer et al. 1986; Williams-Blangero 1993).

Inbred strains of various species, especially rodents, have been developed to address specific research needs (Festing 1979; Gill 1980). When inbred animals or their F1 progeny are used, it is important to periodically monitor genetic authenticity (Festing 1982; Hedrich 1990); several methods of monitoring have been developed that use immunologic, biochemical, and molecular techniques (Cramer 1983; Festing 2002; Groen 1977; Hoffman et al. 1980; Russell et al. 1993). Appropriate management systems (Green 1981; Kempthorne 1957) should be designed to minimize genetic contamination resulting from mutation and mismating.

Genetically modified animals (GMAs) represent an increasingly large proportion of animals used in research and require special consideration in their population management. Integrated or altered genes can interact with species or strain-specific genes, other genetic manipulations, and environmental factors, in part as a function of site of integration, so each GMA line can be considered a unique resource. Care should be taken to preserve such resources through standard genetic management procedures, including maintenance of detailed pedigree records and genetic monitoring to verify the presence and zygosity of transgenes and other genetic modifications (Conner 2005). Cryopreservation of fertilized embryos, ova, ovaries, or spermatozoa should also be considered as a safeguard against alterations in transgenes over time or accidental loss of GMA lines (Conner 2002; Liu et al. 2009).

Generation of animals with multiple genetic alterations often involves crossing different GMA lines and can lead to the production of offspring with genotypes that are not of interest to the researcher (either as experimental or control animals) as well as unexpected phenotypes. Carefully designed breeding strategies and accurate genotype assessment can help to minimize the generation of animals with unwanted genotypes (Linder 2003). Newly generated genotypes should be carefully monitored and new phenotypes that negatively affect well-being should be reported to the IACUC and managed in a manner to ensure the animals' health and well-being.

Accurate recording, with standardized nomenclature when available, of both the strain and substrain or of the genetic background of animals used in a research project is important (NRC 1979b). Several publications provide rules developed by international committees for standardized nomenclature of outbred rodents and rabbits (Festing et al. 1972), inbred rats, inbred mice, and transgenic animals (FELASA 2007; Linder 2003). In addition, the International Committee on Standardized Genetic Nomenclature for Mice and the Rat Genome and Nomenclature Committee maintain online guidelines for these species (MGI 2009).

AQUATIC ANIMALS

The variety of needs for fish and aquatic or semiaquatic reptiles and amphibians is as diverse as the number of species considered. This section is intended to provide facility managers, veterinarians, and IACUCs with basic information related to the management of aquatic animal systems (Alworth and Harvey 2007; Alworth and Vazquez 2009; Browne et al. 2007; Browne and Zippel 2007; Denardo 1995; DeTolla et al. 1995; Koerber and Kalishman 2009; Lawrence 2007; Matthews et al. 2002; Pough 2007). Specific recommendations are available in texts and journal reviews, and it will be necessary to review other literature and consult with experienced caregivers for further detail on caring for aquatic species (see Appendix A).

Aquatic Environment

Microenvironment and Macroenvironment

As with terrestrial systems, the *microenvironment* of an aquatic animal is the physical environment immediately surrounding it—the primary enclosure such as the tank, raceway, or pond. It contains all the resources with which the animals are in direct contact and also provides the limits of the animals' immediate environment. The microenvironment is characterized by many factors, including water quality, illumination, noise, vibration, and temperature. The physical environment of the secondary enclosure, such as a room, constitutes the *macroenvironment*.

Water Quality

The composition of the water (*water quality*) is essential to aquatic animal well-being, although other factors that affect terrestrial microenvironments are also relevant. Water quality parameters and life support systems for aquatic animals will vary with the species, life stage, the total biomass supported, and the animals' intended use (Blaustein et al. 1999; Fisher 2000; Gresens 2004; Overstreet et al. 2000; Schultz and Dawson 2003). The success and adequacy of the system depend on its ability to match the laboratory habitat to the natural history of the species (Godfrey and Sanders 2004; Green 2002; Lawrence 2007; Spence et al. 2008).

Characteristics of the water that may affect its appropriateness include temperature, pH, alkalinity, nitrogen waste products (ammonia, nitrite, and nitrate), phosphorus, chlorine/bromine, oxidation-reduction potential, conductivity/salinity, hardness (osmolality/dissolved minerals), dissolved oxygen, total gas pressure, ion and metal content, and the established microbial ecology of the tank. Water quality parameters can directly affect animal well-being; different classes, species, and ages in a species may have different water quality needs and sensitivities to changes in water quality parameters.

Routine measurement of various water characteristics (water quality testing) is essential for stable husbandry. Standards for acceptable water quality, appropriate parameters to test, and testing frequency should be identified at the institutional level and/or in individual animal use protocols depending on the size of the aquatic program. Staff managing aquatic systems need to be trained in biologically relevant aspects of water chemistry, how water quality parameters may affect animal health and well-being, how to monitor water quality results, and how water quality may affect life support system function (e.g., biologic filtration).

The specific parameters and frequency of testing vary widely (depending on the species, life stage, system, and other factors), from continuous monitoring to infrequent spot checks. Recently established systems and/or populations, or changes in husbandry procedures, may require more frequent assessment as the system ecology stabilizes; stable environments may require less frequent testing. Toxins from system components, particularly in newly constructed systems, may require special consideration such as leaching of chemicals from construction materials, concrete, joint compounds, and sealants (DeTolla et al. 1995; Nickum et al. 2004). Chlorine and chloramines used to disinfect water for human consumption or to disinfect equipment are toxic to fish and amphibians and must be removed or neutralized before use in aquatic systems (Tompkins and Tsai 1976; Wedemeyer 2000).

Life Support System

The phrase *life support system* refers to the physical structure used to contain the water and the animals as well as the ancillary equipment used to move and/or treat the water. Life support systems may be simple (e.g., a container to hold the animal and water) or extremely complex (e.g., a fully automated recirculating system). The type of life support system used depends on several factors including the natural habitat of the species, age/size of the species, number of animals maintained, availability and characteristics of the water required, and the type of research.

Life support systems typically fall into three general categories: recirculating systems where water (all or part) is moved around a system, flow-through systems where water is constantly replaced, or static systems where water is stationary and periodically replenished or replaced. The water may be fresh, brackish, or salt and is maintained at specific temperatures depending on the species' needs.

The source of water for these systems typically falls into four general categories: treated wastewater (e.g., municipal tap water), surface water (e.g., rivers, lakes, or oceans), protected water (e.g., well or aquifer water), or artificial water (e.g., reverse osmosis or distilled water). Artificial saltwater may be created by adding appropriate salt to freshwater sources. Source water selection should be based on the provision of a consistent or constant supply, incoming biosecurity level requirements, water volumes needed, species selection, and research considerations.

Recirculating systems are common in indoor research settings where high-density housing systems are often needed. Most recirculating systems are designed to exchange a specific volume of water per unit time and periodically introduce fresh water into the system. These systems are the most mechanically advanced, containing biologic filters (*biofilters*) that promote conversion of ammonia to nitrite and nitrate via nitrifying bacteria, protein skimmers (foam fractionators) and particulate filters to remove undissolved and dissolved proteins and particulate matter, carbon filters to remove dissolved chemicals, and ultraviolet or ozone units to disinfect the water. The systems generally contain components to aerate and degas the water (to prevent gas oversaturation) and to heat or cool it, as well as automated dosing systems to maintain appropriate pH and conductivity. Not all elements are present in all systems and some components may accomplish multiple functions. Recirculating systems may be designed so that multiple individual tanks are supplied with treated water from a single source, as is the case with "rack" systems used for zebrafish (*Danio rerio*) and *Xenopus laevis* and *X. tropicalis,* as examples (Fisher 2000; Koerber and Kalishman 2009; Schultz and Dawson 2003).

The development and maintenance of the biofilter is critical for limiting ammonia and nitrite accumulation in recirculating systems. The biofilter must be of sufficient size (i.e., contain a sufficient quantity of bacteria) to be capable of processing the bioload (level of nitrogenous waste) entering the system. The microorganisms supported by the biofilter require certain water quality parameters. Alterations in the aquatic environment (e.g., rapid changes in salinity, temperature, and pH) as well as the addition of chemicals or antimicrobials may significantly affect the microbial ecology of the system and therefore water quality and animal well-being. If damaged, biofilter recovery may take weeks (Fisher 2000). Changes in water quality parameters (e.g., pH, ammonia, and nitrite) may negatively affect animal health and the efficiency of the biofilter, so species sensitive to change in water quality outside of a narrow range require more frequent monitoring.

Continuous or timed flow-through systems can be used where suitable water is available to support the species to be housed (e.g., in aquaculture facilities). These systems may use extremely large volumes of water as it is not reused. The water may be used "as is" or processed before use, for example by removing sediments, excessive dissolved gases, chlorine, or chloramines, and by disinfecting with UV or ozone (Fisher 2000; Overstreet et al. 2000). Static systems vary in size from small tanks to large inground ponds, and may use mechanical devices to move and aerate water.

Temperature, Humidity, and Ventilation

The general concepts discussed in the Terrestrial Animals section also apply to the aquatic setting. Most aquatic or semiaquatic species (fish, amphibians, and reptiles) used in research are poikilotherms, which depend, for the most part, on the temperature of their environment to sustain physiologic processes, such as metabolism, reproduction, and feeding behavior (Browne and Edwards 2003; Fraile et al. 1989; Maniero and Carey 1997; Pough 1991). Temperature requirements are based on the natural history of the species and can vary depending on life stage (Green 2002; Pough 1991; Schultz and Dawson 2003). Water temperature may be controlled at its source, within the life support system, or by controlling the macroenvironment. Some semiopen systems (e.g., raceways by a river) depend on source water temperature and thus enclosure water temperature will vary with that of the source water.

The volume of water contained in a room can affect room temperature, temperature stability, and relative humidity. Likewise the thermal load produced by chiller/heater systems can affect the stability of the macroenvironmental temperature. Air handling systems need to be designed to compensate for these thermal and moisture loads. Macroenvironmental relative humidity levels are generally defined by safety issues and staff comfort, since room humidity is not critical for aquatic species; however, excessive moisture may result in condensation on walls, ceilings, and tank lids, which may support microbial growth and serve as a source of contamination or create a conducive environment for metal corrosion. In a dry environment (e.g., indoor heating during cold weather or outdoor housing in some climates/seasons), evaporation rates may be higher, potentially requiring the addition of large quantities of water to the system and monitoring for increases in salinity/conductivity, contaminants, or other water quality aberrations. Some amphibians and reptiles may need elevated microenvironmental humidity (in excess of 50-70% relative humidity), which may require maintaining elevated macroenvironmental humidity levels (Pough 1991; St. Claire et al. 2005).

Room air exchange rates are typically governed by thermal and moisture loads. For fish and some aquatic amphibians, the microenvironmental air quality may affect water quality (i.e., gas exchange), but appropriate life support system design may reduce its importance. Airborne particulates and compounds (e.g., volatile organic compounds and ammonia) may dissolve in tank water and affect animal health (Koerber and Kalishman 2009). As the aerosolization of water can lead to the spread of aquatic animal pathogens (e.g., protozoa, bacteria) within or throughout an aquatic animal facility, this process should be minimized as much as possible (Roberts-Thomson et al. 2006; Wooster and Bowser 2007; Yanong 2003).

Illumination

Aquatic and semiaquatic species are often sensitive to changes in photoperiod, light intensity, and wavelength (Brenner and Brenner 1969). Lighting characteristics will vary by species, their natural history, and the research being conducted. Gradual changes in room light intensity are recommended, as rapid changes in light intensity can elicit a startle response in fish and may result in trauma. Some aquatic and semiaquatic species may need full-spectrum lighting and/or heat lamps to provide supplemental heating to facilitate adequate physiological function (e.g., aquatic turtles provided with a basking area; Pough 1991).

Noise and Vibration

General concepts discussed in the Terrestrial Animals section apply to aquatic animals. These animals may be sensitive to noise and vibration, which are readily transmitted through water. Species vary in their response and many fish species acclimate to noise and vibration, although these may cause subclinical effects (Smith et al. 2007). Vibration through floors can be reduced by using isolation pads under aquaria racks. Some facilities elect to place major components of the life support system (e.g., filters, pumps, and biofilters) outside the animal rooms to reduce vibration and noise.

Aquatic Housing

Microenvironment (Primary Enclosure)

The primary enclosure (a tank, raceway, pond, or pen holding water and the animal) defines the limits of an animal's immediate environment. In research settings, acceptable primary enclosures

- allow for the normal physiological and behavioral needs of the animals, including excretory function, control and maintenance of body temperature, normal movement and postural adjustments, and, where indicated, reproduction. In some poikilothermic reptiles and amphibians, microenvironmental temperature gradients may be needed for certain physiologic functions such as feeding and digestion.
- allow conspecific social interactions (e.g., schooling in fish species).
- provide a balanced, stable environment that supports the animal's physiologic needs.
- provide the appropriate water quality and characteristics, and permit monitoring, filling, refilling, and changing of water.
- allow access to adequate food and removal of food waste.
- restrict escape or accidental entrapment of animals or their appendages.
- are free of sharp edges and/or projections that could cause injury.
- allow for observation of the animals with minimal disturbance.
- are constructed of nontoxic materials that do not leach toxicants or chemicals into the aquatic environment.
- do not present electrical hazards directly or indirectly.

Environmental Enrichment and Social Housing

Environmental enrichment strategies for many aquatic species are not well established. The implications of a barren versus an enriched environment on well-being, general research, growth, and development are unknown or poorly defined, as is true of individual versus group (social) housing for many species. When used, enrichment should elicit species-appropriate behaviors and be evaluated for safety and utility.

Generally, schooling fish species are housed with conspecifics, and many amphibians, especially anuran species, may be group housed. Aggression in aquatic animals does occur (van de Nieuwegiessen et al. 2008; Speedie and Gerlai 2008) and, as for terrestrial animals, appropriate monitoring and intervention may be necessary (Matthews et al. 2002; Torreilles and Green 2007). Some species need appropriate substrate (e.g., gravel) to reproduce or need substrate variety to express basic behaviors and maintain health (Overstreet et al. 2000). Improved breeding success in enriched environments has been reported but further research in this area is needed (Carfagnini et al. 2009). For many species (including, e.g., *X. laevis*), visual barriers, hides, and shading are appropriate (Alworth and Vasquez 2009; Torreilles and Green 2007). Most semiaquatic reptiles spend some time on land (basking, feeding, digesting, and ovipositing) and terrestrial areas should be provided as appropriate.

Sheltered, Outdoor, and Naturalistic Housing

Animals used in aquaculture are often housed in situations that mimic agricultural rearing and may be in outdoor and/or sheltered raceways, ponds, or pens with high population densities. In these settings, where natural predation and mortalities occur, it may be appropriate to measure animal "numbers" by using standard aquaculture techniques such as final production biomass (Borski and Hodson 2003).

Space

Space recommendations and housing density vary extensively with the species, age/size of the animals, life support system, and type of research (Browne et al. 2003; Green 2009; Gresens 2004; Hilken et al. 1995; Matthews et al. 2002). In the United States, for example, adult zebrafish (*Danio rerio*) in typical biomedical research settings are generally housed 5 adult fish per liter of water (Matthews et al. 2002), but this housing density varies when breeding and for housing younger animals (Matthews et al. 2002). This guidance is not necessarily relevant for other species of fish, and may change as research advances (Lawrence 2007). *X. laevis* adults may be housed at 2 liters of water per frog (NRC 1974), but a wide variety of housing systems are currently used in research settings (Green 2009). Institutions, investigators, and IACUC members should evaluate the appropriate needs of each species during program evaluations and facility inspections and continue to review ongoing research in these areas.

Aquatic Management

Behavior and Social Management

Visual evaluations of aquatic and semiaquatic animals are typically used for monitoring. To avoid damage to the protective mucus layers of the skin and negative effects on immune function (De Veer et al. 2007; Subramanian et al. 2007; Tsutsui et al. 2005), handling of these species should be kept to the minimum required (Bly et al. 1997). Appropriate handling techniques vary widely depending on the species, age/size, holding system, and specific research need (Fisher 2000; Matthews et al. 2002; Overstreet et al. 2000); they should be identified at the facility or individual protocol level.

Latex gloves have been associated with toxicity in some amphibians (Gutleb et al. 2001). The use of appropriate nets by well-trained personnel can reduce skin damage and thus stress. Nets should be cleaned and disinfected appropriately when used in different systems and should be dedicated to animals of similar health status whenever possible.

Exercise and activity levels for aquatic species are minimally described but informed decisions may be extrapolated from studies of behavior of the same or similar species in the wild (Spence et al. 2008). Some aquatic species do not rest and constantly swim; others may rest all or a significant portion of the day. Water flow rates and the provision of hides or terrestrial resting platforms (e.g., for some reptiles and amphibians) need to be appropriate for species and life stage.

Husbandry

Food The general principles relating to feeding of terrestrial animals are applicable to aquatic animals. Food should be stored in a type-appropriate manner to preserve nutritional content, minimize contamination, and prevent entry of pests. Food delivery methods should ensure that all animals are able to access food for a sufficient period of time while minimizing feeding aggression and nutrient loss. Feeding methods and frequency vary widely depending on the species, age/size of species, and type of life support system. Many aquatic or semiaquatic species are not provided with food ad libitum in the tank, and in some cases may not be fed daily.

Commercial diets (e.g., pellets, flakes) are available for certain species and storage time should be based on manufacturer recommendations or follow commonly accepted practices. In aquatic systems, particularly in fish rearing or when maintaining some amphibian and reptile species, the use of live foods (e.g., *Artemia* sp. larva, crickets, or mealworm beetle larvae) is common. Live food sources need to be maintained and managed to ensure a steady supply and the health and suitability of the organism as a food. Care should be taken to feed a complete diet to avoid nutritional deficiencies.

Water (see also section on Water Quality) Aquatic animals need access to appropriately conditioned water. Fully aquatic animals obtain water in their habitat or absorb it across their gills or skin. Some semiaquatic amphibians and reptiles may need "bowls" of water for soaking and drinking, and water quality should be appropriate (see Terrestrial Animals section). Chlorine or chloramines may be present in tap water at levels that could be toxic to some species.

Substrate Substrates can provide enrichment for aquatic animals by promoting species-appropriate behavior such as burrowing, foraging, or enhanced spawning (Fisher 2000; Matthews et al. 2002; Overstreet et al. 2000). They may be an integral and essential component of the life support system by providing increased surface area for denitrifying bacteria (e.g., systems with undergravel filtration), and need routine siphoning (i.e., hydrocleaning) to remove organic debris. System design and species needs should be evaluated to determine the amount, type, and presentation of substrate.

Sanitation Sanitation of the aquatic environment in recirculating systems is provided through an appropriately designed and maintained life support system, regular removal of solid waste materials from the enclosure bottom, and periodic water changes. The basic concept of sanitation (i.e., to provide conditions conducive to animal health and welfare) is the same for terrestrial and aquatic systems. However, sanitation measures in aquatic systems differ from those for terrestrial systems because much of the nitrogenous waste (feces and urine) and respiratory output (carbon dioxide) is dissolved in the water.

A properly functioning life support system, designed to process the bioload, will maintain nitrogenous wastes within an acceptable range. Solids may be removed in a variety of ways, depending on the design of the system; generally they are removed by siphoning (hydrocleaning) and/or filtration. Depending on the type, filters need routine cleaning or replacement or, if self-cleaning, proper maintenance; in saltwater systems dissolved proteins may be removed by protein skimmers. Reducing organic solids limits the quantities of nitrogen and phosphorus that need to be removed from the system, both of which can accumulate to levels that are toxic to fish and amphibians. The biologic filter (denitrifying bacteria) typically removes ammonia and nitrite, potential toxins, from aquatic systems. Nitrate, the end product of this process, is less toxic to aquatic animals but at high levels can be problematic; it is generally removed through water changes, although large systems may have a specialized denitrification unit to reduce levels.

Disinfection is usually accomplished through water treatment (e.g., filtration and application of UV light or ozone) and/or water changes. Chlorine and most chemical disinfectants are inappropriate for aquatic systems containing animals as they are toxic at low concentrations; when used to disinfect an entire system or system components, extreme care must be taken to ensure that residual chlorine, chemical, and reactive byproducts are neutralized or removed. The type of monitoring and frequency varies depending on the disinfection method, the system, and the animals.

Algal growth is common in aquatic systems and increases with the presence of nitrogen and phosphorus, particularly in the presence of light. Excessive growth may be an indication of elevated nitrogen or phosphorus levels. Algal species seen with recirculating systems are generally nontoxic, although species capable of producing toxins exist. Algae are typically removed using mechanical methods (i.e., scrubbing or scraping). Limiting algal growth is important to allow viewing of the animals in the enclosure. Cyanobacteria (commonly called blue-green algae) growth is also possible and may be common in freshwater aquaculture. The same factors that promote algae growth also promote cyanobacteria growth. As with algae, while most species are harmless, some species can produce clinically relevant toxic compounds (Smith et al. 2008).

Tank (cage) changing and disinfection are conducted at frequencies using methods that often differ from terrestrial systems. Because waste is dissolved in the water and/or removed as solids by siphoning or filtration, regular changing of tanks is not integral to maintaining adequate hygiene in typical aquatic systems. The frequency of cleaning and disinfection should be determined by water quality, which should permit adequate viewing of the animals, and animal health monitoring. System components such as lids on fish tanks, which may accumulate feed, may require sanitation as often as weekly depending on the frequency and type of feed and the system's design.

Cleaning and Disinfection of the Macroenvironment As with terrestrial systems, all components of the animal facility, including animal rooms and support spaces (e.g., storage areas, cage-washing facilities, corridors, and procedure rooms), should be regularly cleaned and disinfected as appropriate to the circumstances and at a frequency determined by the use of the area and the nature of likely contamination. Cleaning agents should be chosen and used with care to ensure there is no secondary contamination of the aquatic systems.

Cleaning implements should be made of materials that resist corrosion and withstand regular sanitation. They should be assigned to specific areas and should not be transported between areas with different risks of contamination without prior disinfection. Worn items should be replaced regularly. The implements should be stored in a neat, organized fashion that facilitates drying and minimizes contamination or harborage of vermin.

Waste Disposal Wastewater treatment and disposal may be necessary in some facilities depending on water volume, quality, and chemical constituents. Local regulations may limit or control the release of wastewater.

Pest Control Terrestrial animal pest control principles apply to aquatic systems but, due to transcutaneous absorption, aquatic and semiaquatic species may be more sensitive to commonly used pest control agents than terrestrial animals. Before use, an appropriate review of chemicals and methods of application is necessary.

Emergency, Weekend, and Holiday Care As with terrestrial species, aquatic animals should receive daily care from qualified personnel who have a sufficient understanding of the housing system to identify malfunctions and, if they are unable to address a system failure of such magnitude that it requires resolution before the next workday, access to staff who can respond to the problem. Automated monitoring systems are available and may be appropriate depending on system size and complexity. Appropriate emergency response plans should be developed to address major system failures.

Population Management

Identification Identification principles are similar to those for terrestrial animals. Identification criteria are based on the species and housing system. Identification methods available for use in aquatic species include fin clipping, genetic testing (Matthews et al. 2002; Nickum et al. 2004), identification tags, subcutaneous injections of elastomeric or other materials (Nickum et al. 2004), individual transponder tags (in animals of sufficient size), and, as applicable, external features such as individual color patterns. Because it can be difficult to individually identify some small aquatic animals throughout their life, group identification may be more appropriate in some situations (Koerber and Kalishman 2009; Matthews et al. 2002).

Aquatic Animal Recordkeeping Adequate recordkeeping is necessary in aquatic system management. In general, the same standards used for terrestrial animals apply to aquatic and semiaquatic species, although modifications may be necessary to account for species or system variations (Koerber and Kalishman 2009).

Although many aquatic animals are maintained using group (vs. individual) identification, detailed animal records are still necessary. Animal information that may routinely be captured, particularly in biomedical research with fish, includes species; genetic information (parental stock identification, genetic composition); stock source; stock numbers in system; tank identification; system life support information; breeding; deaths;

illnesses; animal transfers within and out of the facility; and fertilization/hatching information (Koerber and Kalishman 2009; Matthews et al. 2002). Records should be kept concerning feeding information (e.g., food offered, acceptance), nonexpired food supplies to ensure sustenance of nutritional profile, and any live cultures (e.g., hatch rates and information to ensure suppliers' recommendations are being met; Matthews et al. 2002).

Records of water quality testing for system and source water and maintenance activities of the life support system components are important for tracking and ensuring water quality. The exact water quality parameters tested and testing frequency should be clearly established and will vary with such factors as the type of life support system, animals, and research, as discussed under Water Quality. Detailed tracking of animal numbers in aquatic systems is often possible with accurate records of transfers, breeding, and mortalities (Matthews et al. 2002). In some cases where animals are housed in large groups (e.g., some *Xenopus* colonies) periodic censuses may be undertaken to obtain an exact count. In large-scale aquaculture research it may be more appropriate to measure biomass of the system versus actual numbers of animals (Borski and Hodson 2003).

REFERENCES

Alworth LC, Harvey SB. 2007. IACUC issues associated with amphibian research. ILAR J 48:278-289.

Alworth LC, Vazquez VM. 2009. A novel system for individually housing bullfrogs. Lab Anim 38:329-333.

Ames BN, Shigenaga MK, Hagen TM. 1993. Review: Oxidants, antioxidants, and the degenerative diseases of aging. Proc Natl Acad Sci USA 90:7915-7922.

Anadon A, Martinez-Larranaga MR, Martinez MA. 2009. Use and abuse of pyrethrins and synthetic pyrethroids in veterinary medicine. Vet J (UK) 182:7-20.

Andrade CS, Guimaraes FS. 2003. Anxiolytic-like effect of group housing on stress-induced behavior in rats. Depress Anx 18:149-152.

Apeldoorn EJ, Schrama JW, Mashaly MM, Parmentier HK. 1999. Effect of melatonin and lighting schedule on energy metabolism in broiler chickens. Poultry Sci 78:223-229.

Arakawa H. 2005. Age dependent effects of space limitation and social tension on open-field behavior in male rats. Physiol Behav 84:429-436.

Armario A, Castellanos JM, Balasch J. 1985. Chronic noise stress and insulin secretion in male rats. Physiol Behav 34:359-361.

Armstrong KR, Clark TR, Peterson MR. 1998. Use of corn-husk nesting material to reduce aggression in caged mice. Contemp Top Lab Anim Sci 37:64-66.

Augustsson H, Lindberg L, Hoglund AU, Dahlborn K. 2002. Human-animal interactions and animal welfare in conventionally and pen-housed rats. Lab Anim 36:271-281.

Azar TA, Sharp JL, Larson DM. 2008. Effect of housing rats in dim light or long nights on heart rate. JAALAS 47:25-34.

Baer LA, Corbin BJ, Vasques MF, Grindeland RE. 1997. Effects of the use of filtered microisolator tops on cage microenvironment and growth rate of mice. 1997. Lab Anim Sci 47:327-329.

Baldwin AL. 2007. Effects of noise on rodent physiology. Int J Comp Psychol 20:134-144.

Barnett SA. 1965. Adaptation of mice to cold. Biol Rev 40:5-51.

Barnett SA. 1973. Maternal processes in the cold-adaptation of mice. Biol Rev 48:477-508.

Bartolomucci A, Palanza P, Parmigiani S. 2002. Group housed mice: Are they really stressed? Ethol Ecol Evol 14:341-350.

Bartolomucci A, Palanza P, Sacerdote P, Ceresini G, Chirieleison A, Panera AE, Parmigiani S. 2003. Individual housing induces altered immuno-endocrine responses to psychological stress in male mice. Psychoneuroendocrinology 28:540-558.

Baumans V. 1997. Environmental enrichment: Practical applications. In: Van Zutphen LFM, Balls M, eds. Animal Alternatives, Welfare and Ethics. Elsevier. p 187-197.

Baumans V. 2005. Environmental enrichment for laboratory rodents and rabbits: Requirements of rodents, rabbits, and research. ILAR J 46:162-170.

Baumans V, Schlingmann F, Vonck M, Van Lith HA. 2002. Individually ventilated cages: Beneficial for mice and men? Contemp Top Lab Anim Sci 41:13-19.

Bayne KA. 2002. Development of the human-research animal bond and its impact on animal well-being. ILAR J 43:4-9.

Bayne KA. 2005. Potential for unintended consequences of environmental enrichment for laboratory animals and research results. ILAR J 46:129-139.

Bayne KA, Dexter SL, Hurst JK, Strange GM, Hill EE. 1993. Kong Toys for laboratory primates: Are they really an enrichment or just fomites? Lab Anim Sci 43(1):78-85.

Bayne KA, Haines MC, Dexter SL, Woodman D, Evans C. 1995. Nonhuman primate wounding prevalence: A retrospective analysis. Lab Anim 24:40-43.

Bazille PG, Walden SD, Koniar BL, Gunther R. 2001. Commercial cotton nesting material as a predisposing factor for conjunctivitis in athymic nude mice. Lab Anim (NY) 30:40-42.

Beaumont S. 2002. Ocular disorders of pet mice and rats. Vet Clin North Am Exot Anim Pract 5:311-324.

Becker BA, Christenson RK, Ford JJ, Nienaber JA, DeShazer JA, Hahn GL. 1989. Adrenal and behavioral responses of swine restricted to varying degrees of mobility. Physiol Behav 45:1171-1176.

Bell GC. 2008. Optimizing laboratory ventilation rates. Labs for the 21st century: Best practice guide. US Environmental Protection Agency. Available at http://labs21century.gov/pdf/bp_opt_vent_508.pdf; accessed March 30, 2010.

Bellhorn RW. 1980. Lighting in the animal environment. Lab Anim Sci 30:440-450.

Bergmann P, Militzer K, Büttner D. 1994. Environmental enrichment and aggressive behaviour: influence on body weight and body fat in male inbred HLG mice. J Exp Anim Sci 37:59-78.

Berson DM, Dunn FA, Takao M. 2002. Phototransduction by retinal ganglion cells that set the circadian clock. Science 295:1070-1073.

Besch EL. 1980. Environmental quality within animal facilities. Lab Anim Sci 30:385-406.

Blaustein A, Marco A, Quichano C. 1999. Sensitivity to nitrate and nitrite in pond-breeding amphibians from the Pacific Northwest, USA. Environ Toxicol Chem J 18:2836-2839.

Blom HJM, Van Tintelen G, Van Vorstenbosch CJ, Baumans V, Beynen AC. 1996. Preferences of mice and rats for types of bedding material. Lab Anim 30:234-244.

Bloomsmith MA, Stone AM, Laule GE. 1998. Positive reinforcement training to enhance the voluntary movement of group-housed chimpanzees within their enclosures. Zoo Biol 17:333-341.

Bly JE, Quiniou SM, Clem LW. 1997. Environmental effects on fish immune mechanisms. Dev Biol Stand 90:33-43.

Borski R, Hodson RG. 2003. Fish research and the institutional animal care and use committee. ILAR J 44:286-294.

Bracke MBM, Metz JHM, Spruijt BM, Schouten WGP. 2002. Decision support system for overall welfare assessment in pregnant sows. B: Validation by expert opinion. J Anim Sci 80:1835-1845.

Brainard GC. 1989. Illumination of laboratory animal quarters: Participation of light irradiance and wavelength in the regulation of the neuroendocrine system. In: Science and Animals: Addressing Contemporary Issues. Greenbelt, MD: Scientists Center for Animal Welfare. p 69-74.

Brainard GC, Vaughan MK, Reiter RJ. 1986. Effect of light irradiance and wavelength on the Syrian hamster reproductive system. Endocrinology 119:648-654.

Brenner FJ, Brenner PE. 1969. The influence of light and temperature on body fat and reproductive conditions of *Rana pipiens*. Ohio J Sci 69:305-312.

Brent L. 1995. Feeding enrichment and body weight in captive chimpanzees. J Med Primatol 24(1):12-16.

Briese V, Fanghanel J, Gasow H. 1984. Effect of pure sound and vibration on the embryonic development of the mouse. Zentralbl Gynokol 106:378-388.

Broderson JR, Lindsey JR, Crawford JE. 1976. The role of environmental ammonia in respiratory mycoplasmosis of rats. Am J Pathol 85:115-127.

Brown AM, Pye JD. 1975. Auditory sensitivity at high frequencies in mammals. Adv Comp Physiol Biochem 6:1-73.

Browne RK, Edwards DL. 2003. The effect of temperature on the growth and development of green and golden bell frogs (*Litoria aurea*). J Therm Biol 28:295-299.

Browne RK, Zippel K. 2007. Reproduction and larval rearing of amphibians. ILAR J 48:214-234.

Browne RK, Pomering M, Hamer AJ. 2003. High density effects on the growth, development and survival of *Litoria aurea* tadpoles. Aquaculture 215:109-121.

Browne RK, Odum RA, Herman T, Zippel K. 2007. Facility design and associated services for the study of amphibians. ILAR J 48:188-202.

Buddaraju AKV, Van Dyke RW. 2003. Effect of animal bedding on rat liver endosome acidification. Comp Med 53:616-621.

Carfagnini AG, Rodd FH, Jeffers KB, Bruce AEE. 2009. The effects of habitat complexity on aggression and fecundity in zebrafish (*Danio rerio*). Environ Biol Fish 86:403-409.

Carissimi AS, Chaguri LCAA, Teixeira MA, Mori CMC, Macchione M, Sant'Anna ETG, Saldiva PHN, Souza NL, Merusse JBL. 2000. Effects of two ventilation systems and bedding change frequency on cage environmental factors in rats (*Rattus norvegicus*). Anim Tech 51:161-170.

Carman RA, Quimby FW, Glickman GM. 2007. The effect of vibration on pregnant laboratory mice. Noise-Con Proc 209:1722-1731.

Castelhano-Carlos MJ, Sousa N, Ohl F, Baumans V. 2010. Identification methods in newborn C57BL/6 mice: A developmental and behavioural evaluation. Lab Anim 4:88-103.

Caulfield CD, Cassidy JP, Kelly JP. 2008. Effects of gamma irradiation and pasteurization on the nutritive composition of commercially available animal diets. JAALAS 47:61-66.

CFR [Code of Federal Regulations]. 2009. Title 21, Part 58. Good Laboratory Practice for Nonclinical Laboratory Studies. Washington: Government Printing Office. Available at www.accessdata.fda.gov/scripts/cdrh/cfdocs/cfcfr/CFRSearch.cfm?CFRPart=58andshowFR=1; accessed April 1, 2010.

Chapillon P, Manneche C, Belzung C, Caston J. 1999. Rearing environmental enrichment in two inbred strain of mice: 1. Effects on emotional reactivity. Behav Genet 29:41-46.

Cherry JA. 1987. The effect of photoperiod on development of sexual behavior and fertility in golden hamsters. Physiol Behav 39:521-526.

Chmiel DJ, Noonan M. 1996. Preference of laboratory rats for potentially enriching stimulus objects. Lab Anim 30:97-101.

Clarence WM, Scott JP, Dorris MC, Paré M. 2006. Use of enclosures with functional vertical space by captive rhesus monkeys (*Macaca mulatta*) involved in biomedical research. JAALAS 45:31-34.

Clough G. 1982. Environmental effects on animals used in biomedical research. Biol Rev 57:487-523.

Colman RJ, Anderson RM, Johnson SC, Kastman EK, Kosmatka KJ, Beasley TM, Allison DB, Cruzen C, Simmons HA, Kemnitz JW, Weindruch R. 2009. Caloric restriction delays disease onset and mortality in rhesus monkeys. Science 325:201-204.

Compton SR, Homberger FR, MacArthur Clark J. 2004a. Microbiological monitoring in individually ventilated cage systems. Lab Anim 33:36-41.

Compton SR, Homberger FR, Paturzo FX, MacArthur Clark J. 2004b. Efficacy of three microbiological monitoring methods in a ventilated cage rack. Comp Med 54:382-392.

Conner DA. 2002. Mouse colony management. Curr Protoc Mol Biol 23.8.1-23.8.11, suppl 57.

Conner DA. 2005. Transgenic mouse colony management. Curr Protoc Mol Biol 23.10.1-23.10.8, suppl 71.

Corning BF, Lipman NS. 1991. A comparison of rodent caging system based on microenvironmental parameters. Lab Anim Sci 41:498-503.

Cramer DV. 1983. Genetic monitoring techniques in rats. ILAR News 26:15-19.

Crippa L, Gobbi A, Ceruti RM. 2000. Ringtail in suckling Munich Wistar Frömter rats: A histopathologic study. Comp Med 50:536-539.

Cunliffe-Beamer TL, Freeman LC, Myers DD. 1981. Barbiturate sleep time in mice exposed to autoclaved or unautoclaved wood beddings. Lab Anim Sci 31:672-675.

Davidson LP, Chedester AL, Cole MN. 2007. Effects of cage density on behavior in young adult mice. Comp Med 57:355-359.

De Boer SF, Koolhaas JM. 2003. Defensive burying in rodents: Ethology, neurobiology and psychopharmacology. Eur J Pharmacol 463:145-161.

De Veer MJ, Kemp JM, Meeusen ENT. 2007. The innate host defence against nematode parasites. Parasite Immunol 29:1-9.

Denardo D. 1995. Amphibians as laboratory animals. ILAR J 37:173-181.

DeTolla LJ, Sriniva S, Whitaker BR, Andrews C, Hecker B, Kane AS, Reimschuessel R. 1995. Guidelines for the care and use of fish in research. ILAR J 37:159-172.

DHHS [Department of Health and Human Services]. 2009. Biosafety in Microbiological and Biomedical Laboratories, 5th ed. Chosewood LC, Wilson DE, eds. Washington: Government Printing Office. Available at http://www.cdc.gov/biosafety/publications/bmbl5/index.htm; accessed July 30, 2010.

DiBerardinis L, Greenley P, Labosky M. 2009. Laboratory air changes: What is all the hot air about? J Chem Health Safety 16:7-13.

Donahue WA, VanGundy DN, Satterfield WC, Coghlan LJ. 1989. Solving a tough problem. Pest Control 57:46-50.

Drescher B. 1993. The effects of housing systems for rabbits with special reference to ulcerative pododermatitis. Tierärztl Umschau 48:72-78.

Duncan TE, O'Steen WK. 1985. The diurnal susceptibility of rat retinal photoreceptors to light-induced damage. Exp Eye Res 41:497-507.

Dyke B. 1993. Basic data standards for primate colonies. Am J Primatol 29:125-143.

Eadie JM, Mann SO. 1970. Development of the rumen microbial population: High starch diets and instability. In: Phillipson AT, Annison EF, Armstrong DG, Balch CC, Comline RS, Hardy RN, Hobson PN, Keynes RD, eds. Physiology of Digestion and Metabolism in the Ruminant. Proceedings of the Third International Symposium. Newcastle upon Tyne UK: FRS Oriel Press Ltd. p 335-347.

Easterbrook JD, Kaplan JB, Glass GE, Watson J, Klein SL. 2008. A survey of rodent-borne pathogens carried by wild-caught Norway rats: A potential threat to laboratory rodent colonies. Lab Anim 42:92-98.

Ednie DL, Wilson RP, Lang CM. 1998. Comparison of two sanitation monitoring methods in an animal research facility. Contemp Top Lab Anim Sci 37:71-74.

Erkert HG, Grober J. 1986. Direct modulation of activity and body temperature of owl monkeys (*Aotus lemurinus griseimembra*) by low light intensities. Folia Primatol 47:171-188.

Eskola S, Lauhikari M, Voipio HM, Laitinen M, Nevalainen T. 1999. Environmental enrichment may alter the number of rats needed to achieve statistical significance. Scand J Lab Anim Sci 26:134-144.

FELASA [Federation of European Laboratory Animal Science Associations] Working Group. 2007. FELASA Guidelines for the production and nomenclature of transgenic rodents. Lab Anim 41:301-311.

Festing MFW. 1979. Inbred Strains in Biomedical Research. London: Macmillan.

Festing MFW. 1982. Genetic contamination of laboratory animal colonies: An increasingly serious problem. ILAR News 25:6-10.

Festing MFW. 2002. Laboratory animal genetics and genetic quality control. In: Hau J, Van Hoosier GL Jr, eds. Handbook of Laboratory Animal Science. Boca Raton, FL: CRC Press. p 173-203.

Festing MFW, Kondo K, Loosli R, Poiley SM, Spiegel A. 1972. International standardized nomenclature for outbred stocks of laboratory animals. ICLA Bull 30:4-17.

Fidler IJ. 1977. Depression of macrophages in mice drinking hyperchlorinated water. Nature 270:735-736.

Field K, Bailey M, Foresman LL, Harris RL, Motzel SL, Rockar RA, Ruble G, Suckow MA. 2007. Medical records for animals used in research, teaching and testing: Public statement from the American College of Laboratory Animal Medicine. ILAR J 48:37-41.

Fisher JP. 2000. Facilities and husbandry (large fish model). In: Ostrander GK, ed. The Laboratory Fish. San Francisco: Academic Press. p 13-39.

Fletcher JL. 1976. Influence of noise on animals. In: McSheehy T, ed. Control of the Animal House Environment. Laboratory Animal Handbooks 7. London: Laboratory Animals Ltd. p 51-62.

Fraile B, Paniagua R, Rodrigues MC, Saez J. 1989. Effects of photoperiod and temperature on spermiogenesis in marbeled newts (*Triturus marmoratus marmoratus*). Copeia 1989:357-363.

Fullerton FR, Greenman DL, Kendall DC. 1982. Effects of storage conditions on nutritional qualities of semipurified (AIN-76) and natural ingredient (NIH-07) diets. J Nutr 112:567-573.

Fullerton PM, Gilliatt RW. 1967. Pressure neuropathy in the hind foot of the guinea pig. J Neurol Neurosurg Psychiat 30:18-25.

Garg RC, Donahue WA. 1989. Pharmacologic profile of methoprene, an insect growth regulator, in cattle, dogs, and cats. JAVMA 194:410-412.

Garner JP. 2005. Stereotypies and other abnormal repetitive behaviors: Potential impact on validity, reliability, and replicability of scientific outcomes. ILAR J 46:106-117.

Garrard G, Harrison GA, Weiner JS. 1974. Reproduction and survival of mice at 23°C. J Reprod Fert 37:287-298.

Gärtner K. 1999. Cage enrichment occasionally increases deviation of quantitative traits. In: Proc Int Joint Mtg 12th ICLAS General Assembly and Conference and 7th FELASA Symposium. p 207-210.

Gaskill BN, Rohr SA, Pajor EA, Lucas JR, Garner JP. 2009. Some like it hot: Mouse temperature preferences in laboratory housing. Appl Anim Behav Sci 116:279-285.

Geber WF, Anderson TA, Van Dyne B. 1966. Physiologic responses of the albino rat to chronic noise stress. Arch Environ Health 12:751-754.

Georgsson L, Barrett J, Gietzen D. 2001. The effects of group-housing and relative weight on feeding behaviour in rats. Scand J Lab Anim Sci 28:201-209.

Gibson SV, Besch-Williford C, Raisbeck MF, Wagner JE, McLaughlin RM. 1987. Organophosphate toxicity in rats associated with contaminated bedding. Lab Anim 37:789-791.

Gill TJ. 1980. The use of randomly bred and genetically defined animals in biomedical research. Am J Pathol 101(3S):S21-S32.

Godfrey EW, Sanders GW. 2004. Effect of water hardness on oocyte quality and embryo development in the African clawed frog (*Xenopus laevis*). Comp Med 54:170-175.

Gonder JC, Laber K. 2007. A renewed look at laboratory rodent housing and management. ILAR J 48:29-36.

Gonzalez RR, Kiuger MJ, Hardy JD. 1971. Partitional calorimetry of the New Zealand white rabbit at temperatures of 5-35°C. J Appl Physiol 31:728.

Gordon AH, Hart PD, Young MR. 1980. Ammonia inhibits phagosome-lysosome fusion in macrophages. Nature 286:79-80.

Gordon CJ. 1990. Thermal biology of the laboratory rat. Physiol Behav 47:963-991.

Gordon CJ. 1993. Temperature Regulation in Laboratory Animals. New York: Cambridge University Press.

Gordon CJ. 2004. Effect of cage bedding on temperature regulation and metabolism of group-housed female mice. Comp Med 54:63-68.

Gordon CJ. 2005. Temperature and Toxicology: An integrative, comparative and environmental approach. Boca Raton, FL: CRC Press.

Gordon CJ, Becker P, Ali JS. 1998. Behavioral thermoregulatory responses of single- and group-housed mice. Physiol Behav 65:255-262.

Green EL. 1981. Genetics and Probability in Animal Breeding Experiments. New York: Oxford University Press.

Green SL. 2002. Factors affecting oogenesis in the South African clawed frog (*Xenopus laevis*). Comp Med 52:307-312.

Green SL. 2009. The Laboratory *Xenopus* sp. (Laboratory Animal Pocket Reference). Boca Raton, FL: CRC Press.

Greenman DL, Bryant P, Kodell RL, Sheldon W. 1982. Influence of cage shelf level on retinal atrophy in mice. Lab Anim Sci 32:353-356.

Gresens J. 2004. An introduction to the Mexican axolotl (*Ambystoma mexicanum*). Lab Anim 33:41-47.

Groen A. 1977. Identification and genetic monitoring of mouse inbred strains using biomedical polymorphisms. Lab Anim (London) II:209-214.

Gunasekara AS, Rubin AL, Goh KS, Spurlock FC, Tjeerdema RS. 2008. Environmental fate and toxicology of carbaryl. Rev Environ Contam Toxicol 196:95-121.

Gutleb AC, Bronkhorst M, van den Berg JHJ, Musrk AJ. 2001. Latex laboratory gloves: An unexpected pitfall in amphibian toxicity assays with tadpoles. Environ Toxicol Pharmacol 10:119-121.

Haemisch A, Voss T, Gärtner K. 1994 Effects of environmental enrichment on aggressive behaviour, dominance hierarchies and endocrine states in male DBA/2J mice. Physiol Behav 56:1041-1048.

Hahn NE, Lau D, Eckert K, Markowitz H. 2000. Environmental enrichment-related injury in a macaque (*Macaca fascicularis*): Intestinal linear foreign body. Comp Med 50:556-558.

Hall FS. 1998. Social deprivation of neonatal, adolescent, and adult rats has distinct neurochemical and behavioural consequences. Crit Rev Neurobiol 12:129-162.

Hall JE, White WJ, Lang CM. 1980. Acidification of drinking water: Its effects on selected biologic phenomena in male mice. Lab Anim Sci 30:643-651.

Hankenson FC, Garzel LM, Fischer DD, Nolan B, Hankenson KD. 2008. Evaluation of tail biopsy collection in laboratory mice (*Mus musculus*): Vertebral ossification, DNA quantity, and acute behavioral responses. JAALAS 47(6):10-18.

Hanifin JP, Brainard GC. 2007. Photoreception for circadian, neuroendocrine, and neurobehavioral regulation. J Physiol Anthropol 26:87-94.

Hartl DL. 2000. A Primer of Population Genetics, 3rd ed. Sunderland, MA: Sinauer Associates.

Hasenau JJ, Baggs RB, Kraus AL. 1993. Microenvironments in microisolation cages using BALB/c and CD-1 Mice. Contemp Top Lab Anim Sci 32:11-16.

Hedrich HJ. 1990. Genetic Monitoring of Inbred Strains of Rats. New York: Gustav Fischer Verlag.

Heffner HE, Heffner RS. 2007. Hearing ranges of laboratory animals. JAALAS 46:20-22.

Held SDE, Turner RJ, Wootton RJ. 1995. Choices of laboratory rabbits for individual or group-housing. Appl Anim Behav Sci 46:81-91

Hermann LM, White WJ, Lang CM. 1982. Prolonged exposure to acid, chlorine, or tetracycline in drinking water: Effects on delayed-type hypersensitivity, hemagglutination titers, and reticuloendothelial clearance rates in mice. Lab Anim Sci 32:603-608.

Hess SE, Rohr S, Dufour BD, Gaskill BN, Pajor EA, Garner JP. 2008. Home improvement: C57BL/6J mice given more naturalistic nesting materials build better nests. JAALAS 47:25-31.

Hilken G, Dimigen J, Iglauer F. 1995. Growth of *Xenopus laevis* under different laboratory rearing conditions. Lab Anim 29:152-162.

Hill D. 1999. Safe handling and disposal of laboratory animal waste. Occup Med 14:449-468.

Hoffman HA, Smith KT, Crowell JS, Nomura T, Tomita T. 1980. Genetic quality control of laboratory animals with emphasis on genetic monitoring. In: Spiegel A, Erichsen S, Solleveld HA, eds. Animal Quality and Models in Biomedical Research. Stuttgart: Gustav Fischer Verlag. p 307-317.

Homberger FR, Pataki Z, Thomann PE. 1993. Control of *Pseudomonas aeruginosa* infection in mice by chlorine treatment of drinking water. Lab Anim Sci 43:635-637.

Hotchkiss CE, Paule MG. 2003. Effect of pair-housing on operant behavior task performance by rhesus monkeys. Contemp Top Lab Anim Sci 42:38-41.

Hubrecht RC. 1993. A comparison of social and environmental enrichment methods for laboratory housed dogs. Appl Anim Behav Sci 37:345-361.

Hughes HC, Reynolds S. 1995. The use of computational fluid dynamics for modeling air flow design in a kennel facility. Contemp Top Lab Anim Sci 34:49-53

Ikemoto S, Panksepp J. 1992. The effect of early social isolation on the motivation for social play in juvenile rats. Dev Psychobiol 25:261-274.

Ivy AS, Brunson KL, Sandman C, Baram TZ. 2008. Dysfunctional nurturing behavior in rat dams with limited access to nesting material: A clinically relevant model for early-life stress. Neuroscience 154:1132-1142.

Jacobs BB, Dieter DK. 1978. Spontaneous hepatomas in mice inbred from Ha:ICR Swiss stock: Effects of sex, cedar shavings in bedding, and immunization with fetal liver or hepatoma cells. J Natl Cancer Inst 61:1531-1534.

Jones DM. 1977. The occurrence of dieldrin in sawdust used as bedding material. Lab Anim 11:137.

Karolewicz B, Paul IA. 2001. Group housing of mice increases immobility and antidepressant sensitivity in the forced swim and tail suspension tests. Eur J Pharmacol 415:97-201.

Kaufman BM, Pouliot AL, Tiefenbacher S, Novak MA. 2004. Short- and long-term effects of a substantial change in cage size on individually housed, adult male rhesus monkeys (*Macaca mulatta*). Appl Anim Behav Sci 88:319-330.

Kaye GI, Weber PB, Evans A, Venezia RA. 1998. Efficacy of alkaline hydrolysis as an alternative method for treatment and disposal of infectious animal waste. Contemp Top Lab Anim Sci 37:43-46.

Keenan KP, Smith PF, Soper KA. 1994. Effect of dietary (caloric) restriction on aging, survival, pathobiology and toxicology. In: Notter W, Dungworth DL, Capen CC, eds. Pathobiology of the Aging Rat, vol 2. Washington: International Life Sciences Institute. p 609-628.

Keenan KP, Laroque P, Ballam GC, Soper KA, Dixit R, Mattson BA, Adams SP, Coleman JB. 1996. The effects of diet, ad libitum overfeeding, and moderate dietary restriction on the rodent bioassay: The uncontrolled variable in safety assessment. Toxicol Pathol 24:757-768.

Keller LSF, White WJ, Snyder MT, Lang CM. 1989. An evaluation of intracage ventilation in three animal caging systems. Lab Anim Sci 39:237-242.

Kempthorne O. 1957. An Introduction to Genetic Statistics. New York: John Wiley and Sons.

King JE, Bennett GW. 1989. Comparative activity of fenoxycarb and hydroprene in sterilizing the German cockroach (Dictyoptera: Blattellidae). J Econ Entomol 82:833-838.

Knapka JJ. 1983. Nutrition. In: Foster HL, Small JD, Fox JG, eds. The Mouse in Biomedical Research, vol III: Normative Biology, Immunology and Husbandry. New York: Academic Press. p 52-67.

Koerber AS, Kalishman J. 2009. Preparing for a semi-annual IACUC inspection of a satellite zebrafish (*Danio rerio*) facility. JAALAS 48:65-75.

Kraft LM. 1980. The manufacture, shipping and receiving, and quality control of rodent bedding materials. Lab Anim Sci 30:366-376.

Krause J, McDonnell G, Riedesel H. 2001. Biodecontamination of animal rooms and heat-sensitive equipment with vaporized hydrogen peroxide. Contemp Top Lab Anim Sci 40: 8-21.

Krohn TC, Hansen AK, Dragsted N. 2003. The impact of cage ventilation on rats housed in IVC systems. Lab Anim 37:85-93.

Laber K, Veatch L, Lopez M, Lathers D. 2008. The impact of housing density on weight gain, immune function, behavior, and plasma corticosterone levels in BALB/c and C57Bl/6 mice. JAALAS 47:6-23.

Lacy RC. 1989. Analysis of founder representation in pedigrees: Founder equivalents and founder genome equivalents. Zoo Biol 8:111-123.

Lanum J. 1979. The damaging effects of light on the retina: Empirical findings, theoretical and practical implications. Surv Ophthalmol 22:221-249.

Laule GE, Bloomsmith MA, Schapiro SJ. 2003 The use of positive reinforcement training techniques to enhance the care, management, and welfare of primates in the laboratory. J Appl Anim Welf Sci 6:163-173.

Lawler DF, Larson BT, Ballam JM, Smith GK, Biery DN, Evan RH, Greeley EH, Segre M, Stowe HD, Kealy RD. 2008. Diet restriction and ageing in the dog: Major observations over two decades. Br J Nutr 99:793-805.

Lawlor MM. 2002. Comfortable quarters for rats in research institutions. In: Reinhardt V, Reinhardt A, eds. Comfortable Quarters for Laboratory Animals, 9th ed. Washington: Animal Welfare Institute. p 26-32.

Lawrence C. 2007. The husbandry of zebrafish (*Danio rerio*): A review. Aquaculture 269:1-20.

Leveille GA, Hanson RW. 1966. Adaptive changes in enzyme activity and metabolic pathways in adipose tissue from meal-fed rats. J Lipid Res 7:46.

Linder CC. 2003. Mouse nomenclature and maintenance in genetically engineered mice. Comp Med 53:119-125.

Lipman NS. 1993. Strategies for architectural integration of ventilated caging systems. Contemp Top Lab Anim Sci 32:7-10.

Liu L, Nutter LMJ, Law N, McKerlie C. 2009. Sperm freezing and in vitro fertilization on three substrains of C57BL/6 mice. JAALAS 48:39-43.
Lupo C, Fontani G, Girolami L, Lodi L, Muscettola M. 2000. Immune and endocrine aspects of physical and social environmental variations in groups of male rabbits in seminatural conditions. Ethol Ecol Evol 12:281-289.
Lutz CK, Novak MA. 2005. Environmental enrichment for nonhuman primates: Theory and application. ILAR J 46:178-191.
MacCluer JW, VandeBerg JL, Read B, Ryder OA. 1986. Pedigree analysis by computer simulation. Zoo Biol 5:147-160.
MacLean EL, Prior RS, Platt ML, Brannon EM. 2009. Primate location preference in a double-tier cage: The effects of illumination and cage height. J Anim Welf Sci 12:73-81.
Macrì S, Pasquali P, Bonsignore LT, Pieretti S, Cirulli F, Chiarotti F, Laviola G. 2007. Moderate neonatal stress decreases within-group variation in behavioral, immune and HPA responses in adult mice. PLoS One 2(10):e1015.
Maniero GD, Carey C. 1997. Changes in selected aspects of immune function in leopard frog, *Rana pipiens*, associated with exposure to cold. J Comp Physiol B 167:256-263.
Manninen AS, Antilla S, Savolainen H. 1998. Rat metabolic adaptation to ammonia inhalation. Proc Soc Biol Med 187:278-281.
Manser CE, Morris TH, Broom DM. 1995. An investigation into the effects of solid or grid cage flooring on the welfare of laboratory rats. Lab Anim 29:353-363.
Manser CE, Elliott H, Morris TH, Broom DM. 1996. The use of a novel operant test to determine the strength of preference for flooring in laboratory rats. Lab Anim 30:1-6.
Manser CE, Broom DM, Overend P, Morris TM. 1997. Operant studies to determine the strength of preference in laboratory rats for nest boxes and nest materials. Lab Anim 32:36-41.
Manser CE, Broom DM, Overend P, Morris TM. 1998. Investigations into the preferences of laboratory rats for nest boxes and nesting materials. Lab Anim 32:23-35.
Martin B, Ji S, Maudsley S, Mattson MP. 2010. "Control" laboratory rodents are metabolically morbid: Why it matters. Proc Nat Acad Sci USA 107:6127-6133.
Mason G, Littin KE. 2003. The humaneness of rodent pest control. Anim Welf 12:1-37.
Matthews M, Trevarrow B, Matthews J. 2002. A virtual tour of the guide for zebrafish users. Lab Anim 31:34-40.
McCune S. 1997. Enriching the environment of the laboratory cat: A review. In: Proceedings of the Second International Conference on Environmental Enrichment, August 21-25, 1995, Copenhagen Zoo, Denmark. p 103-117.
McGlone JJ, Anderson DL, Norman RL. 2001. Floor space needs for laboratory mice: BALB/cJ males or females in solid-bottom cages with bedding. Contemp Top Lab Anim Sci 40:21-25.
Meerburg BG, Brom FWA, Kijlstra A. 2008. The ethics of rodent control. Pest Manag Sci 64:1205-1211.
Meier TR, Maute CJ, Cadillac JM, Lee JY, Righter DJ, Hugunin KMS, Deininger RA, Dysko RC. 2008. Quantification, distribution, and possible source of bacterial biofilm in mouse automated watering systems. JAVMA 42:63-70.
Memarzadeh F, Harrison PC, Riskowski GL, Henze T. 2004. Comparisons of environment and mice in static and mechanically ventilated isolator cages with different air velocities and ventilation designs. Contemp Top Lab Anim Sci 43:14-20.
MGI [Mouse Genome Informatics]. 2009. Guidelines for Nomenclature of Genes, Genetic Markers, Alleles, and Mutations in Mouse and Rat. International Committee on Standardized Genetic Nomenclature for Mice and Rat Genome and Nomenclature Committee. Available at www.informatics.jax.org/mgihome/nomen/gene.shtml; accessed May 10, 2010.
Moore BJ. 1987. The California diet: An inappropriate tool for studies of thermogenesis. J Nutrit 117:227-231.

Murphy RGL, Scanga JA, Powers BE, Pilon JL, VerCauteren KC, Nash PB, Smith GC, Belk KE. 2009. Alkaline hydrolysis of mouse-adapted scrapie for inactivation and disposal of prion-positive material. J Anim Sci 87:1787-1793.

Nadelkov M. 1996. EPA impact on pathological incineration: What will it take to comply? Lab Anim 25:35-38.

NAFA [National Air Filtration Association]. 1996. NAFA Guide to Air Filtration, 2nd ed. Virginia Beach.

NASA [National Aeronautics and Space Administration]. 1988. Summary of conclusions reached in workshop and recommendations for lighting animal housing modules used in microgravity related projects. In: Holley DC, Winget CM, Leon HA, eds. Lighting Requirements in Microgravity: Rodents and Nonhuman Primates. NASA Technical Memorandum 101077. Moffett Field, CA: Ames Research Center. p 5-8.

Nayfield KC, Besch EL. 1981. Comparative responses of rabbits and rats to elevated noise. Lab Anim Sci 31:386-390.

Nevalainen T, Vartiainen T. 1996. Volatile organic compounds in commonly used beddings before and after autoclaving. Scand J Lab Anim Sci 23:101-104.

Newberne PM. 1975. Influence on pharmacological experiments of chemicals and other factors in diets of laboratory animals. Fed Proc 34:209-218.

Newberry RC. 1995. Environmental enrichment: Increasing the biological relevance of captive environments. Appl Anim Beh Sci 44:229-243.

Newbold JA, Chapin LT, Zinn SA, Tucker HA. 1991. Effects of photoperiod on mammary development and concentration of hormones in serum of pregnant dairy heifers. J Dairy Sci 74:100-108.

Nickum JG, Bart HL Jr, Bowser PR. 2004. Guidelines for the Use of Fishes in Research. Bethesda, MD: American Fisheries Society.

Njaa LR, Utne F, Braekkan OR. 1957. Effect of relative humidity on rat breeding and ringtail. Nature 180:290-291.

Novak MA, Meyer JS, Lutz C, Tiefenbacher S. 2006. Deprived environments: Developmental insights from primatology. In: Mason G, Rushen J, eds. Stereotypic Animal Behaviour: Fundamentals and Applications to Welfare. Wallingford, UK: CABI. p 153-189.

Novak MF, Kenney C, Suomi SJ, Ruppenthal GC. 2007. Use of animal-operated folding perches by rhesus macaques (*Macaca mulatta*). JAALAS 46:35-43.

NRC [National Research Council]. 1974. Amphibians: Guidelines for the Breeding, Care and Management of Laboratory Animals. Washington: National Academy of Sciences.

NRC. 1977. Nutrient Requirements of Rabbits, 2nd rev ed. Washington: National Academy Press.

NRC. 1979a. Laboratory Animal Records. Washington: National Academy Press.

NRC. 1979b. Laboratory animal management: Genetics. ILAR News 23(1):A1-A16.

NRC. 1982. Nutrient Requirements of Mink and Foxes, 2nd rev ed. Washington: National Academy Press.

NRC. 1989. Biosafety in the Laboratory: Prudent Practices for the Handling and Disposal of Infectious Materials. Washington: National Academy Press.

NRC. 1993. Nutrient Requirements of Fish. Washington: National Academy Press.

NRC. 1994. Nutrient Requirements of Poultry, 9th rev ed. Washington: National Academy Press.

NRC. 1995a. Nutrient Requirements of Laboratory Animals, 4th rev ed. Washington: National Academy Press.

NRC. 1995b. Prudent Practices in the Laboratory: Handling and Disposal of Chemicals. Washington: National Academy Press.

NRC. 1996. Laboratory Animal Management: Rodents. Washington: National Academy Press.

NRC. 1998a. Psychological Well-being of Nonhuman Primates. Washington: National Academy Press.

NRC. 1998b. Nutrient Requirements of Swine, 10th rev ed. Washington: National Academy Press.

NRC. 2000. Nutrient Requirements of Beef Cattle, 7th rev ed: Update 2000. Washington: National Academy Press.

NRC. 2001. Nutrient Requirements of Dairy Cattle, 7th rev ed. Washington: National Academy Press.

NRC. 2003a. Nutrient Requirements of Nonhuman Primates, 2nd rev ed. Washington: National Academies Press.

NRC. 2003b. Guidelines for the Care and Use of Mammals in Neuroscience and Behavioral Research. Washington: National Academies Press.

NRC. 2006a. Preparation of Animals for Use in the Laboratory. ILAR J 43:281-375.

NRC. 2006b. Nutrient Requirements of Dogs and Cats. Washington: National Academies Press.

NRC. 2006c. Nutrient Requirements of Horses, 6th rev ed. Washington: National Academies Press.

NRC. 2007. Nutrient Requirements of Small Ruminants: Sheep, Goats, Cervids, and New World Camelids. Washington: National Academies Press.

Olivier B, Molewijk E, van Oorschot R, van der Poel G, Zethof T, van der Heyden J, Mos J. 1994. New animal models of anxiety. Eur Neuropsychopharmacol 4:93-102.

Olson LC, Palotay JL. 1983. Epistaxis and bullae in cynomolgus macaques (*Macaca fascicularis*). Lab Anim Sci 33:377-379.

Olsson IA, Dahlborn, K. 2002. Improving housing conditions for laboratory mice: A review of "environmental enrichment." Lab Anim 36:243-270.

OSHA [Occupational Safety and Health Administration]. 1998. Occupational Safety and Health Standards. Subpart G, Occupational Health and Environmental Controls, Occupational Noise Exposure (29 CFR 1910.95). Washington: Department of Labor.

O'Steen WK. 1980. Hormonal influences in retinal photodamage. In: Williams TP, Baker BN, eds. The Effects of Constant Light on Visual Processes. New York: Plenum Press. p 29-49.

Overall KL, Dyer D. 2005. Enrichment strategies for laboratory animals from the viewpoint of clinical behavioural veterinary medicine: Emphasis on cats and dogs. ILAR J 46:202-215.

Overstreet RM, Barnes SS, Manning CS, Hawkins W. 2000. Facilities and husbandry (small fish model). In: Ostrander GK, ed. The Laboratory Fish. San Francisco: Academic Press. p 41-63.

Parker A, Wilfred A, Hidell T. 2003. Environmental monitoring: The key to effective sanitation. Lab Anim 32:26-29.

Peace TA, Singer AW, Niemuth NA, Shaw ME. 2001. Effects of caging type and animal source on the development of foot lesions in Sprague-Dawley rats (*Rattus norvegicus*). Contemp Top Lab Anim Sci 40:17-21.

Pekrul D. 1991. Noise control. In: Ruys T, ed. Handbook of Facilities Planning, vol 2: Laboratory Animal Facilities. New York: Van Nostrand Reinhold. p 166-173.

Pennycuik PR. 1967. A comparison of the effects of a range of high environmental temperatures and of two different periods of acclimatization on the reproductive performances of male and female mice. Aust J Exp Biol Med Sci 45:527-532.

Perez C, Canal JR, Dominguez E, Campillo JE, Guillen M, Torres MD. 1997. Individual housing influences certain biochemical parameters in the rat. Lab Anim 31:357-361.

Perkins SE, Lipman NS. 1995. Characterization and qualification of microenvironmental contaminants in isolator cages with a variety of contact bedding. Contemp Top Lab Anim Sci 34:93-98.

Peterson EA. 1980. Noise and laboratory animals. Lab Anim Sci 30:422-439.

Peterson EA, Augenstein JS, Tanis DC, Augenstein DG. 1981. Noise raises blood pressure without impairing auditory sensitivity. Science 211:1450-1452.

Pfaff J, Stecker M. 1976. Loudness levels and frequency content of noise in the animal house. Lab Anim 10:111-117.

Poiley SM. 1960. A systematic method of breeder rotation for non-inbred laboratory animal colonies. Proc Anim Care Panel 10:159-166.

Poole T. 1998. Meeting a mammal's psychological needs. In: Shepherdson DJ, Mellen JD, Hutchins M, eds. Second Nature: Environmental Enrichment for Captive Animals. Washington: Smithsonian Institute Press. p 83-94.

Pough FH. 1991. Recommendations for the care of amphibians and reptiles in academic institutions. ILAR J 33:1-16.

Pough FH. 2007. Amphibian biology and husbandry. ILAR J 48:203-213.

Raje S. 1997. Group housing for male New Zealand white rabbits. Lab Anim 26:36-38.

Ras T, van de Ven M, Patterson-Kane EG, Nelson K. 2002. Rats' preferences for corn versus wood-based bedding and nesting materials. Lab Anim 36:420-425.

Rasmussen S, Glickman G, Norinsky R, Quimby F, Tolwani RJ. 2009. Construction noise decreases reproductive efficiency in mice. JAALAS 48:263-270.

Reeb CK, Jones R, Bearg D, Bedigan H, Myers D, Paigen B. 1998. Microenvironment in ventilated cages with differing ventilation rates, mice populations and frequency of bedding changes. JAALAS 37:70-74.

Reeb-Whitaker CK, Paigen B, Beamer WG, Bronson RT, Churchill GA, Schweitzer IB, Myers DD. 2001. The impact of reduced frequency of cage changes on the health of mice housed in ventilated cages. Lab Anim 35:58-73.

Reinhardt V. 1997. Training nonhuman primates to cooperate during handling procedures: A review. Anim Tech 48:55-73.

Reme CE, Wirz-Justice A, Terman M. 1991. The visual input stage of the mammalian circadian pacemaking system. I: Is there a clock in the mammalian eye? J Biol Rhythms 6:5-29.

Rennie AE, Buchanan-Smith HM. 2006. Refinement of the use of non-human primates in scientific research. Part I: The influence of humans. Anim Welf 15:203-213.

Richmond JY, Hill RH, Weyant RS, Nesby-O'Dell SL, Vinson PE. 2003. What's hot in animal biosafety? ILAR J 44:20-27.

Roberts-Thomson A, Barnes A, Filder DS, Lester RJG, Adlard RD. 2006. Aerosol dispersal of the fish pathogen *Amyloodinium ocellatum*. Aquaculture 257:118-123.

Rock FM, Landi MS, Hughes HC, Gagnon RC. 1997. Effects of caging type and group size on selected physiologic variables in rats. Contemp Top Lab Anim Sci 36:69-72.

Rollin BE. 1990. Ethics and research animals: Theory and practice. In: Rollin B, Kesel M, eds. The Experimental Animal in Biomedical Research, vol I: A Survey of Scientific and Ethical Issues for Investigators. Boca Raton, FL: CRC Press. p 19-36.

Rommers J, Meijerhof R. 1996. The effect of different floor types on foot pad injuries of rabbit does. In: Proceedings of the 6th World Rabbit Science Congress 1996, Toulouse. p 431-436.

Russell RJ, Festing MFW, Deeny AA, Peters AG. 1993. DNA fingerprinting for genetic monitoring of inbred laboratory rats and mice. Lab Anim Sci 43:460-465.

Sales GD. 1991. The effect of 22 kHz calls and artificial 38 kHz signals on activity in rats. Behav Proc 24:83-93.

Sales GD, Milligan SR, Khirnykh K. 1999. Sources of sound in the laboratory animal environment: A survey of the sounds produced by procedures and equipment. Anim Welf 8:97-115.

Saltarelli DG, Coppola CP. 1979. Influence of visible light on organ weights of mice. Lab Anim Sci 29:319-322.

Sanford AN, Clark SE, Talham G, Sidelsky MG, Coffin SE. 2002. Influence of bedding type on mucosal immune responses. Comp Med 52:429-432.

Schaefer DC, Asner IN, Seifert B, Bürki K, Cinelli P. 2010. Analysis of physiological and behavioural parameters in mice after toe clipping as newborns. Lab Anim 44:7-13.

Schlingmann F, De Rijk SHLM, Pereboom WJ, Remie R. 1993a. Avoidance as a behavioural parameter in the determination of distress amongst albino and pigmented rats at various light intensities. Anim Tech 44:87-107.

Schlingmann F, Pereboom W, Remie R. 1993b.The sensitivity of albino and pigmented rats to light. Anim Tech 44:71-85.

Schoeb TR, Davidson MK, Lindsey JR. 1982. Intracage ammonia promotes growth of mycoplasma pulmonis in the respiratory tract of rats. Infect Immun 38:212-217.

Schondelmeyer CW, Dillehay DL, Webb SK, Huerkamp MJ, Mook DM, Pullium JK. 2006. Investigation of appropriate sanitization frequency for rodent caging accessories: Evidence supporting less-frequent cleaning. JAALAS 45:40-43.

Schultz TW, Dawson DA. 2003. Housing and husbandry of *Xenopus* for oocyte production. Lab Anim 32:34-39.

Semple-Rowland SL, Dawson WW. 1987. Retinal cyclic light damage threshold for albino rats. Lab Anim Sci 37:289-298.

Sherwin CM. 2002. Comfortable quarters for mice in research institutions. In: Reinhardt V, Reinhardt A, eds. Comfortable Quarters for Laboratory Animals, 9th ed. Washington: Animal Welfare Institute. p 6-17.

Smith AL, Mabus SL, Stockwell JD, Muir C. 2004. Effects of housing density and cage floor space on C57BL/6J mice. Comp Med 54:656-663.

Smith AL, Mabus SL, Muir C, Woo Y. 2005. Effect of housing density and cage floor space on three strains of young adult inbred mice. Comp Med 55:368-376.

Smith E, Stockwell JD, Schweitzer I, Langley SH, Smith AL. 2004. Evaluation of cage microenvironment of mice housed on various types of bedding materials. Contemp Top Lab Anim Sci 43:12-17.

Smith JL, Boyer GL, Zimba PV. 2008. A review of cyanobacterial odorous and bioactive metabolites: Impacts and management alternatives in aquaculture. Aquaculture 280:5-20.

Smith ME, Kane AD, Popper AN. 2007. Noise-induced stress responsive and hearing loss in goldfish (*Carassius auratus*). J Exp Biol 207:427-435.

Speedie N, Gerlai R. 2008. Alarm substance induced behavioral responses in zebrafish (*Danio rerio*). Behav Brain Res 188:168-177.

Spence R, Gerlach G, Lawrence C, Smith C. 2008. The behavior and ecology of the zebrafish, *Danio rerio*. Biol Rev 83:13-34.

St. Claire MB, Kennett MJ, Thomas ML, Daly JW. 2005. The husbandry and care of dendrobatid frogs. Contemp Top Lab Anim Sci 44:8-14.

Stauffacher M. 1992. Group housing and enrichment cages for breeding, fattening and laboratory rabbits. Anim Welf 1:105-125.

Stoskopf MK. 1983. The physiological effects of psychological stress. Zoo Biol 2:179-190.

Subramanian S, MacKinnon SL, Ross NW. 2007. A comparative study on innate immune parameters in the epidermal mucus of various fish species. Comp Biochem Physiol B Biochem Mol Biol 148:256-263.

Suckow MA, Doerning BJ. 2007. Assessment of veterinary care. In: Silverman J, Suckow MA, Murthy S, eds. The IACUC Handbook, 2nd ed.. Boca Raton, FL: CRC Press.

Terman M, Reme CE, Wirz-Justice A. 1991. The visual input stage of the mammalian circadian pacemaking sytem II: The effect of light and drugs on retinal function. J Biol Rhythms 6:31-48.

Thigpen JE, Lebetkin EH, Dawes ML, Clark JL, Langley CL, Amy HL, Crawford D. 1989. A standard procedure for measuring rodent bedding particle size and dust content. Lab Anim Sci 39:60-62.

Thigpen JE, Setchell KDR, Ahlmark KB, Locklear J, Spahr T, Caviness GF, Goelz MF, Haseman JK, Newbold RR, Forsythe DB. 1999. Phytoestrogen content of purified, open- and closed-formula laboratory animal diets. Lab Anim Sci 49:530-539.

Thigpen JE, Setchell KDR, Saunders HE, Haseman JK, Grant MG, Forsythe DB. 2004. Selecting the appropriate rodent diet for endocrine disruptor research and testing studies. ILAR J 45:401-416.

Tompkins JA, Tsai C. 1976. Survival time and lethal exposure time for the blacknose dace exposed to free chlorine and chloramines. Trans Am Fish Soc 105:313-321.

Torreilles SL, Green SL. 2007. Refuge cover decreases the incidence of bite wounds in laboratory South African clawed frogs (*Xenopus laevis*). JAALAS 46:33-36.

Torronen R, Pelkonen K, Karenlampi S. 1989. Enzyme-inducing and cytotoxic effects of wood-based materials used as bedding for laboratory animals: Comparison by a cell culture study. Life Sci 45:559-565.

Totten M. 1958. Ringtail in newborn Norway rats: A study of the effect of environmental temperature and humidity on incidence. J Hygiene 56:190-196.

Tsai PP, Stelzer HD, Hedrich HJ, Hackbarth H. 2003. Are the effects of different enrichment designs on the physiology and behaviour of DBA/2 mice consistent? Lab Anim 37:314-327.

Tsutsui S, Tasumi S, Suetake H, Kikuchi K, Suzuki Y. 2005. Demonstration of the mucosal lectins in the epithelial cells of internal and external body surface tissues in pufferfish (*Fugu rubripes*). Dev Comp Immun 29:243-253.

Tucker HA, Petitclerc D, Zinn SA. 1984. The influence of photoperiod on body weight gain, body composition, nutrient intake and hormone secretion. J Anim Sci 59:1610-1620.

Turner JG, Bauer CA, Rybak LP. 2007. Noise in animal facilities: Why it matters. JAALAS 46:10-13.

Turner RJ, Held SD, Hirst JE, Billinghurst G, Wootton RJ. 1997. An immunological assessment of group-housed rabbits. Lab Anim 31:362-372.

Twaddle NC, Churchwell MI, McDaniel LP, Doerge DR. 2004. Autoclave sterilization produces acrylamide in rodent diets: Implications for toxicity testing. J Agric Food Chem 52:4344-4349.

USDA [US Department of Agriculture]. 1985. 9 CFR 1A. (Title 9, Chapter 1, Subchapter A): Animal Welfare. Available at http://ecfr.gpoaccess.gov/cgi/t/text/text-idx?sid=8314313bd7adf2c9f1964e2d82a88d92andc=ecfrandtpl=/ecfrbrowse/Title09/9cfrv1_02.tpl; accessed January 14, 2010.

van de Nieuwegiessen PG, Boerlage AS, Verreth JAJ, Schrama AW. 2008. Assessing the effects of a chronic stressor, stocking density, on welfare indicators of juvenile African catfish, *Clarias gariepinus* Burchell. Appl Anim Behav Sci 115:233-243.

van den Bos R, de Cock Buning T. 1994. Social behaviour of domestic cats (*Felis lybica catus* L.): A study of dominance in a group of female laboratory cats. Ethology 98:14-37.

Van Loo PL, Mol JA, Koolhaas JM, Van Zutphen BM, Baumans V. 2001. Modulation of aggression in male mice: Influence of group size and cage size. Physiol Behav 72:675-683.

Van Loo PL, Van de Weerd HA, Van Zutphen LF, Baumans V. 2004. Preference for social contact versus environmental enrichment in male laboratory mice. Lab Anim 38:178-188.

van Praag H, Kempermann G, Gage FH. 2000. Neural consequences of environmental enrichment. Nat Rev Neurosci 1:191-198.

Verma RK. 2002. Advances on cockroach control. Asian J Microbiol, Biotech Environ Sci 4:245-249.

Vesell ES. 1967. Induction of drug-metabolizing enzymes in liver microsomes of mice and rats by softwood bedding. Science 157:1057-1058.

Vesell ES, Lang CM, White WJ, Passananti GT, Tripp SL. 1973. Hepatic drug metabolism in rats: Impairment in a dirty environment. Science 179:896-897.

Vesell ES, Lang CM, White WJ, Passananti GT, Hill RN, Clemen TL, Liu DL, Johnson WD. 1976. Environmental and genetic factors affecting response of laboratory animals to drugs. Fed Proc 35:1125-1132.

Vlahakis G. 1977. Possible carcinogenic effects of cedar shavings in bedding of C3H-A^{vy}fB mice. J Natl Cancer Inst 58:149-150.

Vogelweid CM. 1998. Developing emergency management plans for university laboratory animal programs and facilities. Contemp Top Lab Anim Sci 37:52-56.

Waiblinger E. 2002. Comfortable quarters for gerbils in research institutions. In: Reinhardt V, Reinhardt A, eds. Comfortable Quarters for Laboratory Animals, 9th ed. Washington: Animal Welfare Institute. p 18-25.

Wardrip CL, Artwohl JE, Bennett BT. 1994. A review of the role of temperature versus time in an effective cage sanitation program. Contemp Top Lab Anim Sci 33:66-68.

Wardrip CL, Artwohl JE, Oswald J, Bennett BT. 2000. Verification of bacterial killing effects of cage wash time and temperature combinations using standard penicylinder methods. Contemp Top Lab Anim Sci 39:9-12.

Wax TM. 1977. Effects of age, strain, and illumination intensity on activity and self-selection of light-dark schedules in mice. J Comp Physiol Psychol 91:51-62.

Wedemeyer GA. 2000. Chlorination/dechlorination. In: Stickney RR, ed. Encyclopaedia of Aquaculture. Chichester: John Wiley and Sons. p 172-174.

Weed JL, Watson LM. 1998. Pair housing adult owl monkeys (*Aotus* sp.) for environmental enrichment. Am J Primatol 45:212.

Weichbrod RH, Hall JE, Simmonds RC, Cisar CF. 1986. Selecting bedding material. Lab Anim 15:25-29.

Weichbrod RH, Cisar CF, Miller JG, Simmonds RC, Alvares AP, Ueng TH. 1988. Effects of cage beddings on microsomal oxidative enzymes in rat liver. Lab Anim Sci 38:296-298.

Weihe WH. 1971. Behavioural thermoregulation in mice with change of cooling power of the air. Int J Biometeorol 15:356-361.

Weindruch R, Walford RL. 1988. The Retardation of Aging and Disease by Dietary Restriction. Springfield, IL: Charles C Thomas.

White WJ, Hawk CT, Vasbinder MA. 2008. The use of laboratory animals in toxicology research. In: Hays AW, ed. Principles and Methods in Toxicology, 5th ed. Boca Raton, FL: CRC Press. p 1055-1101.

Williams LE, Steadman A, Kyser B. 2000. Increased cage size affects *Aotus* time budgets and partner distances. Am J Primatol 51(Suppl 1):98.

Williams-Blangero S. 1991. Recent trends in genetic research on captive and wild nonhuman primate populations. Yearb Phys Anthropol 34:69-96.

Williams-Blangero S. 1993. Research-oriented genetic management of nonhuman primate colonies. Lab Anim Sci 43:535-540.

Willott JF. 2007. Factors affecting hearing in mice, rats, and other laboratory animals. JAALAS 46:23-27.

Wolfensohn S. 2004. Social housing of large primates: Methodology for refinement of husbandry and management. Altern Lab Anim 32(Suppl 1A):149-151.

Wolfer DP, Litvin O, Morf S, Nitsch RM, Lipp HP, Würbel H. 2004. Laboratory animal welfare: Cage enrichment and mouse behaviour. Nature 432:821-822.

Wolff A, Rupert G. 1991. A practical assessment of a nonhuman primate exercise program. Lab Anim 20:36-39.

Wooster GA, Bowser PR. 2007. The aerobiological pathway of a fish pathogen: Survival and disseminaton of *Aeromonas salmonicida* in aerosols and its implications in fish health management. J World Aquacul Soc 27:7-14.

Würbel H. 2001. Ideal homes? Housing effects on rodent brain and behaviour. Trends Neurosci 24:207-211.

Yanong RPE. 2003. Fish health management considerations in recirculating aquaculture systems, part 2: Pathogens. IFAS, University of Florida. Available at www.aces.edu/dept/fisheries/aquaculture/documents/fishhealth2.pdf; accessed April 15, 2010.

Yildiz A, Hayirli A, Okumus Z, Kaynar O, Kisa F. 2007. Physiological profile of juvenile rats: Effects of cage size and cage density. Lab Anim 36:28-38.

Young RJ. 2003. Environmental Enrichment for Captive Animals. UFAW Animal Welfare Series. London: Blackwell Science.

4

Veterinary Care

Veterinary care is an essential part of an animal care and use Program. The primary focus of the veterinarian is to oversee the well-being and clinical care of animals used in research, testing, teaching, and production. This responsibility extends to monitoring and promoting animal well-being at all times during animal use and during all phases of the animal's life. Well-being is determined by considering physical, physiologic, and behavioral indicators, which vary by species. The number, species, and use of animals housed in an institution may influence the complexity of the veterinary care program, but a veterinary program that offers a high quality of care and ethical standards must be provided, regardless of the number of animals or species maintained.

An adequate veterinary care program consists of assessment of animal well-being and effective management of

- animal procurement and transportation
- preventive medicine (including quarantine, animal biosecurity, and surveillance)
- clinical disease, disability, or related health issues
- protocol-associated disease, disability, and other sequelae
- surgery and perioperative care
- pain and distress
- anesthesia and analgesia
- euthanasia.

The veterinary care program is the responsibility of the attending veterinarian (AV), who is certified or has training or experience in laboratory animal science and medicine or is otherwise qualified in the care of the species being used. Some aspects of the veterinary care program can be conducted by persons other than a veterinarian, but a mechanism for direct and frequent communication should be established to ensure that timely and accurate information is conveyed to the responsible veterinarian about issues associated with animal health, behavior, and well-being, and that appropriate treatment or euthanasia is administered. The AV should provide guidance to investigators and all personnel involved in the care and use of animals to ensure appropriate husbandry, handling, medical treatment, immobilization, sedation, analgesia, anesthesia, and euthanasia. In addition, the AV should provide guidance and oversight to surgery programs and perioperative care involving animals.

ANIMAL PROCUREMENT AND TRANSPORTATION

Animal Procurement

All animals must be acquired lawfully, and the receiving institution should ensure that all procedures involving animal procurement are conducted in a lawful manner. Before procuring animals, the principal investigator should confirm that there are sufficient facilities and expertise to house and manage the species being acquired. Procurement of animals should be linked to the prior approval of animal use and number by the IACUC (see Chapter 2, Protocol Review). If dogs and cats are obtained from random sources, such as shelters or pounds, the animals should be inspected for tattoos or identification devices such as subcutaneous transponders (NRC 2009b); such identification might indicate that an animal was a pet, and if so, ownership should be verified. Attention should also be given to the population status of the species under consideration; the threatened or endangered status of species is updated annually by the Fish and Wildlife Service (DOI 2007). Appropriate records and other forms of documentation should be maintained for animals acquired by an institution for its investigators.

Potential vendors should be evaluated for the quality of animals they supply. As a rule, vendors of purpose-bred animals (e.g., USDA Class A dealers) regularly provide information that describes the genetic and pathogen status of their colonies or individual animals and relevant clinical history (e.g., vaccination status and anthelminthic administration). The use of purpose-bred and preconditioned animals is therefore preferable when consistent with the research, teaching, and testing objectives. In general, animals used for scientific purposes should not be obtained from pet stores or pet distributors due to the unknown or uncontrolled background of animals from these sources and the potential for introducing health risks to personnel and other facility animals. Breeding colonies should be established based on need and managed according to principles of animal reduction such as cryopreservation for rodent stocks or strains (Robinson et al. 2003).

Transportation of Animals

Transportation of animals is governed by a number of US regulatory agencies and international bodies. The Animal Welfare Regulations (USDA 1985) set standards for interstate and export/import transportation of regulated species; the International Air Transport Association (IATA) updates the Live Animals Regulations annually and IATA member airlines and many countries agree to comply with these regulations to ensure the safe and humane transport of animals by air (IATA 2009). The Centers for Disease Control and Prevention and USDA enforce regulations to prevent the introduction, transmission, or spread of communicable diseases and regulate the importation of any animal or animal product capable of carrying a zoonotic disease. The US Fish and Wildlife Service regulates importation/exportation of wild vertebrate and invertebrate animals and their tissues. As the national authority arm of the Convention on International Trade in Endangered Species of Wild Fauna and Flora (CITES), the US Fish and Wildlife Service also regulates movement of CITES-listed species that are captive bred, including nonhuman primates (DOI 2007). Institutions should contact appropriate authorities to ensure compliance with any relevant statutes and other animal transportation requirements that must be met for animals to cross international boundaries, including those not of the country of final destination. The NRC publication *Guidelines for the Humane Transportation of Research Animals* provides a comprehensive review of this topic (NRC 2006); additional references on transportation of animals are available in Appendix A.

Animal transportation may be intrainstitutional, interinstitutional, or between a commercial or noncommercial source and a research facility. For wildlife, transportation may occur between the capture site and field holding facilities. Careful planning for all types of transportation should occur to ensure animal safety and well-being. The process of transportation should provide an appropriate level of animal biosecurity (see definition on page 109) while minimizing zoonotic risks, protecting against environmental extremes, avoiding overcrowding, providing for the animals' physical, physiologic, or behavioral needs and comfort, and protecting the animals and personnel from physical trauma (Maher and Schub 2004).

Movement of animals within or between sites or institutions should be planned and coordinated by responsible and well-trained persons at the sending and receiving sites to minimize animal transit time or delays in receipt. Shipping should be coordinated to ensure that animals arrive during normal business hours or, if delivery occurs outside of this time, that someone is available to receive them. Defining and delegating responsibility to the appropriate persons, who are knowledgeable about the needs of the species being shipped, will help ensure effective communication and planning of animal transport (AVMA 2002).

All animals in transit within and between institutions or jurisdictions should be accompanied by appropriate documentation to minimize delays in shipping and receipt. Documentation may include health certificates, sending and receiving institutions' addresses and contacts, emergency procedures and veterinary contact information, and agency permits as needed.

For noncommercial sources of animals, in particular, it is important for the veterinarian or the veterinarian's designee to review the health status and other housing and husbandry requirements before authorizing shipment of animals. This action will ensure that effective quarantine practices are implemented for incoming animals and address any special requirements needed to ensure animal well-being (Otto and Tolwani 2002). Special considerations may be necessary for transporting animals during certain phases of their life or in certain conditions, such as pregnant, perinatal, and geriatric animals; animals with preexisting medical conditions (e.g., diabetes mellitus); and animals surgically prepared by the supplier (FASS 2010).

Although ensuring animal biosecurity during transportation is always important, it is of particular importance for immunocompromised, genetically modified, and specific pathogen-free rodents (Jacoby and Lindsey 1998). For these animals, reinforced disposable shipping containers with filter-protected ventilation openings and internal food and water sources help ensure that microbial contamination does not occur during transit. Commercial vendors are experienced in animal transport and typically use dedicated transport systems and protocols to minimize microbiologic contamination. Noncommercial or interinstitutional transfer of rodents poses a higher risk of microbial contamination since the individuals involved may lack the required knowledge and animal biosecurity capabilities to maintain the animals' health status. Risks due to in-transit microbial contamination of shipping container surfaces can be reduced by decontaminating the surfaces before placement of the containers in clean sites of animal facilities (NRC 1996, 2006). Transportation of animals in private vehicles is discouraged because of potential animal biosecurity, safety, health, and liability risks for the animals, personnel, and institution.

For aquatic species and amphibians, special considerations are required for transportation in an aqueous or sufficiently moist environment, and special attention should be given to avoiding temperature extremes for poikilotherms.

In all cases, appropriate loading and unloading facilities should be provided for the safe and secure transfer of animals at an institution. Facilities and procedures should be in place to help ensure that the environment at the site does not pose risks to animal well-being or personnel safety. During times of extreme temperatures animal transport may be detrimental to animal well-being and therefore may not be possible unless an appropriately heated or cooled means of transportation is available (Robertshaw 2004; Schrama et al. 1996).

PREVENTIVE MEDICINE

Disease prevention is an essential component of comprehensive veterinary medical care and biosecurity programs. Effective preventive medicine enhances the research value of animals by maintaining healthy animals and minimizing nonprotocol sources of variation associated with disease and inapparent infection, thus minimizing animal waste and potential effects on well-being. Preventive medicine programs consist of various combinations of policies, procedures, and equipment related to quarantine and stabilization and the separation of animals by species, source, and health status.

Animal Biosecurity

Animal biosecurity refers to all measures taken to identify, contain, prevent, and eradicate known or unknown infections that may cause clinical disease or alter physiologic and behavioral responses or otherwise make the animals unsuitable for research. Animal biosecurity practices should be applied to all species, but they are most important when housing large numbers of animals in intensive housing conditions (e.g., laboratory rodents). Limiting exposure of animals to infectious disease agents requires consideration of physical plant layout and operational practices. Separation of clean and soiled caging and equipment, and sometimes the associated staff, is often fundamental to success.

> *Animal biosecurity* includes all measures to control known or unknown infections in laboratory animals.

A successful animal biosecurity program incorporates a number of elements: procedures that ensure that only animals of a desired defined health status enter the facility; personnel and materials, especially consumables, that do not serve as fomites; practices that reduce the likelihood of cross contamination if an infectious agent is inadvertently introduced; a comprehensive ongoing system for evaluating animals' health status, including access to all animals; and containment and eradication, if desired, of introduced infectious agents. Related components include procedures for evaluating and selecting appropriate animal suppliers (these may include quarantine and determination of animal health status if unknown); treatment of animals or their products at entry to minimize disease risks (e.g., surface disinfection of fish eggs); a comprehensive pest control program that may include evaluation of the health status of feral animals; procedures to ensure that all biologics administered to animals are free of contamination; and procedures for intra- and interfacility animal transport (e.g., transport of animals to laboratory and other facilities outside the animal facility can present challenges to animal biosecurity) (Balaban and Hampshire 2001). Additional details pertaining to these topics are available in the sections of Chapter 2 that deal with occupational health and safety.

Quarantine and Stabilization

Quarantine is the separation of newly received animals from those already in the facility, in a way that prevents potential spread of contaminants, until the health and possibly the microbial status of the newly received animals have been determined. Transportation can be stressful and may induce recrudescence of subclinical infections harbored by an animal.

An effective quarantine program minimizes the risk of introduction of pathogens into an established colony. The veterinary medical staff should implement procedures for evaluating the health and, if appropriate, the pathogen status of newly received animals, and the procedures should reflect acceptable veterinary medical practice and federal and state regulations applicable to zoonoses (Butler et al. 1995). Effective quarantine procedures are particularly helpful in limiting human exposure to zoonotic infections from nonhuman primates, such as mycobacterial infections, which necessitate specific guidelines for handling of these animals (Lerche et al. 2008; Roberts and Andrews 2008).

Information from suppliers about animal quality should be sufficient to enable a veterinarian to establish the length of quarantine, define the potential risks to personnel and animals in the colony, determine whether therapy is required before animals are released from quarantine, and, in the case of rodents, determine whether rederivation (cesarean or embryo transfer) is necessary to free the animals of specific pathogens. Rodents may not require quarantine if data from the vendor or provider are sufficiently current, complete, and reliable to define the health status of the incoming animals and if the potential for exposure to pathogens during transit is considered. When quarantine is indicated, animals from one shipment should be handled separately or be physically separated from animals from other shipments to preclude transfer of infectious agents between groups.

Depending on the health status of the colony animals and consistent with the animal biosecurity program in place, rodents or other animals being moved outside an animal facility for procedures (e.g., imaging or behavioral testing) may need to be held separately from their colony of origin until their health status is evaluated.

Regardless of whether the animals are quarantined, newly received animals should be given a period for physiologic, behavioral, and nutritional acclimation before their use (Obernier and Baldwin 2006). The length of time for acclimation will depend on the type and duration of animal transportation, the species, and the intended use of the animals. For animals not typically housed in research settings, consideration should be given to providing means to assist with their acclimation (e.g., shearing sheep before they are brought indoors). The need for an acclimation period has been demonstrated in mice, rats, guinea pigs, nonhuman primates, and goats, and time for acclimation is likely important for other species as well (Capitanio et al. 2006; Conour et al. 2006; Kagira et al. 2007; Landi et al. 1982; Prasad et al. 1978; Sanhouri et al. 1989; Tuli et al. 1995).

Separation by Health Status and Species

Physical separation of animals by species is recommended to prevent interspecies disease transmission and to eliminate the potential for anxiety and physiologic and behavioral changes due to interspecies conflict (Arndt et al. 2010). Such separation is usually accomplished by housing different species in separate rooms, but in some instances it may be possible with cubicles, laminar flow units, cages that have filtered air or separate ventilation, or isolators. It may also be acceptable to house different species in the same room—for example, two species that have a similar pathogen status and are behaviorally compatible (Pritchett-Corning et al. 2009), or aquatic species, as long as nets and other animal handling devices remain separate between systems.

In some species subclinical or latent infections can cause clinical disease if transmitted to another species. A few examples may serve as a guide in determining the need for separate housing by species:

- *Helicobacter bilis* can infect rats and mice and may induce clinical disease in both species (Haines et al. 1998; Jacoby and Lindsey 1998; Maggio-Price et al. 2002).
- As a rule, New World (South and Central American), Old World African, and Old World Asian species of nonhuman primates should be housed in separate rooms. Simian hemorrhagic fever (Renquist 1990) and simian immunodeficiency virus (Hirsch et al. 1991; Murphey-Corb et al. 1986), for example, cause only subclinical

infections in African species but induce clinical disease in Asian species.

- Some species should be housed in separate rooms even though they are from the same geographic region. For example, squirrel monkeys (*Saimiri sciureus*) and tamarins (*Saguinus oedipus*) may be latently infected with herpesviruses (*Herpesvirus saimiri* and *H. tamarinus*, respectively), which could be transmitted to and cause a fatal epizootic disease in owl monkeys (*Aotus trivirgatus*) (Barahona et al. 1975; Hunt and Melendez 1966; Murphy et al. 1971).

Intraspecies separation may be essential when animals obtained from multiple sites or sources, either commercial or institutional, differ in pathogen status—for example, with respect to rat theilovirus in rats, mouse hepatitis virus in mice, bacterial gill disease in rainbow trout, *Pasteurella multocida* in rabbits, *Macacine herpesvirus* 1 (B virus) in macaque species, and *Mycoplasma hyopneumoniae* in swine.

Surveillance, Diagnosis, Treatment, and Control of Disease

All animals should be observed for signs of illness, injury, or abnormal behavior by a person trained to recognize such signs. As a rule, such observation should occur at least daily, but more frequent observations may be required, such as during postoperative recovery, when animals are ill or have a physical deficit, or when animals are approaching a study endpoint. Professional judgment should be used to ensure that the frequency and character of observations minimize risks to individual animals and do not compromise the research for which the animals are used.

Appropriate procedures should be in place for disease surveillance and diagnosis. Unexpected deaths and signs of illness, distress, or other deviations from normal in animals should be reported promptly and investigated, as necessary, to ensure appropriate and timely delivery of veterinary medical care. Animals that show signs of a contagious disease should be isolated from healthy animals. If an entire room or enclosure of animals is known or believed to be exposed to an infectious agent (e.g., *Mycobacterium tuberculosis* in nonhuman primates), the group should be kept intact during the process of diagnosis, treatment, and control.

Procedures for disease prevention, diagnosis, and therapy should be those currently accepted in veterinary and laboratory animal practice. Health monitoring programs also include veterinary herd/flock health programs for livestock and colony health monitoring programs for aquatic and rodent species. Access to diagnostic laboratory services facilitates veterinary medical care and can include gross and microscopic pathology, hematology, microbiology, parasitology, clinical chemistry, molecular diagnostics, and serology. If a disease or infectious agent is identified in a facility or colony, the choice of therapy should be made by the veterinarian in consultation with the investigator. If the animal is to remain in the study, the selected treatment plan should be therapeutically sound and, when possible, interfere minimally with the research process.

Subclinical microbial infections (see Appendix A, Pathology, Clinical Pathology, and Parasitology) occur frequently in conventionally maintained rodents but can also occur in facilities designed and maintained for production and use of pathogen-free rodents if the microbial barrier is breached. Examples of infectious agents that can be subclinical but that may induce immunologic changes or alter physiologic, pharmacologic, or toxicologic responses are noroviruses, parvoviruses, mouse hepatitis virus, lymphocytic choriomeningitis virus, and *Helicobacter* spp. (Besselsen et al. 2008; Clifford and Watson 2008; NRC 1991a,b,c). Scientific objectives of a particular protocol, the consequences of infection in a specific strain of rodent, the potential for zoonotic disease, and the adverse effects that infectious agents may have on other animals or protocols in a facility should determine the characteristics of rodent health surveillance programs and strategies for keeping rodents free of specific pathogens.

The principal methods for detecting microbial infections in animal populations are serologic tests (e.g., flow cytometric bead immunoassays, immunofluorescent assays) but other methods, such as DNA analysis using polymerase chain reaction (PCR), microbial culture, clinical chemistry (e.g., lactate dehydrogenase virus), histopathology, and other validated emerging technologies, can also be used to make or confirm a diagnosis.

Transplantable tumors, hybridomas, cell lines, blood products, and other biologic materials can be sources of both murine and human viruses that can contaminate rodents or pose risks to laboratory personnel (Nicklas et al. 1993); rapid and effective assays are available to monitor microbiologic contamination and should be considered before introducing such material into animals (Peterson 2008).

Because health monitoring programs are dependent on the size and complexity of the Program, the species involved, and the institutional research focus, it is beyond the scope of the *Guide* to go into details about health monitoring programs for all species; additional references are in Appendix A (under Disease Surveillance, Diagnosis, and Treatment; Pathology, Clinical Pathology, and Parasitology; and Species-Specific References).

CLINICAL CARE AND MANAGEMENT

Healthy, well-cared-for animals are a prerequisite for good-quality animal-based science. The structure of the veterinary care program, including the number of qualified veterinarians, should be appropriate to fulfill the program's requirements, which will vary by institution, species used, and the nature of the animal use. To be effective in providing clinical care, the veterinarian should be familiar with the species and various uses of animals in the institutional research, teaching, testing, or production programs and have access to medical and experimental treatment records.

Medical Management

There should be a timely and accurate method for communication of any abnormalities in or concerns about animal health, behavior, and well-being to the veterinarian or the veterinarian's designee. The responsibility for communicating these concerns rests with all those involved with animal care and use. Reports should be triaged to ensure that animals most in need receive priority attention, and the veterinarian or veterinarian's designee should perform an objective assessment of the animal(s) to determine an appropriate course of action.

Well-planned experiments with clearly delineated scientific and humane endpoints will help to ensure that a contingency plan is in place for problems that may arise during the study (see Chapter 2, Experimental and Humane Endpoints). For animals on research protocols, the veterinarian or veterinarian's designee should make every effort to discuss any problems with the principal investigator or project director to jointly determine the most appropriate course of treatment or action. Standard operating procedures (SOPs) may be developed for recurrent health conditions to expedite treatment. Recurrent or significant problems involving experimental animal health should be communicated to the IACUC, and all treatments and outcomes should be documented (USDA 1997).

Emergency Care

Procedures must be in place to provide for emergency veterinary care both during and outside of regularly scheduled hours. Such procedures must enable animal care and research staff to make timely reports of animal injury, illness, or death. A veterinarian or the veterinarian's designee must be available to expeditiously assess the animal's condition, treat the animal, investigate an unexpected death, or advise on euthanasia. In the case of a pressing health problem, if the responsible person (e.g., investigator) is not available or if the investigator and veterinary staff cannot reach consensus on treatment, the veterinarian must have the authority, delegated by senior administration (see Chapter 2, Institutional Official and Attending Veterinarian) and the IACUC, to treat the animal, remove it from the experiment, institute appropriate measures to relieve severe pain or distress, or perform euthanasia if necessary.

Recordkeeping

Medical records are a key element of the veterinary care program and are considered critical for documenting animal well-being as well as tracking animal care and use at a facility. A veterinarian should be involved in establishing, reviewing, and overseeing medical and animal use records (Field et al. 2007; Suckow and Doerning 2007). All those involved in animal care and use must comply with federal laws and regulations regarding human and veterinary drugs and treatments. Drug records and storage procedures should be reviewed during facility inspections.

SURGERY

Successful surgical outcomes require appropriate attention to presurgical planning, personnel training, anesthesia, aseptic and surgical technique, assessment of animal well-being, appropriate use of analgesics, and animal physiologic status during all phases of a protocol involving surgery and postoperative care (see Appendix A, Anesthesia, Pain, and Surgery). The individual impact of those factors will vary according to the complexity of procedures involved and the species of animal used. A team approach to a surgical project often increases the likelihood of a successful outcome by providing input from persons with different expertise (Brown and Schofield 1994; Brown et al. 1993).

Surgical outcomes should be continually and thoroughly assessed to ensure that appropriate procedures are followed and timely corrective changes are instituted. Modification of standard techniques may be required (for instance, in aquatic or field surgery), but should not compromise the well-being of the animals. In the event of modification, close assessment of outcomes may have to incorporate criteria other than clinical morbidity and mortality. Such assessments rely on continuing communication among technical staff, investigators, veterinarians, and the IACUC.

Training

Researchers conducting surgical procedures must have appropriate training to ensure that good surgical technique is practiced—that is, asepsis, gentle tissue handling, minimal dissection of tissue, appropriate use of instruments, effective hemostasis, and correct use of suture materials and patterns (Brown et al. 1993; Heon et al. 2006). Training may have to be tailored to accommodate the wide range of educational backgrounds frequently encountered in research settings. For example, persons trained in human surgery may need training in interspecies variations in anatomy, physiology, the effects of anesthetic and analgesic drugs, and/or postoperative care requirements.

Technical staff performing rodent surgery may have had little formal training in surgical techniques and asepsis and may require general surgical training as well as training for the specific techniques they are expected to perform (Stevens and Dey 2007).

Training guidelines for research surgery commensurate with an individual's background are available (ASR 2009) to assist institutions in developing appropriate training programs. The IACUC, together with the AV, is responsible for determining that personnel performing surgical procedures are appropriately qualified and trained in the procedures (Anderson 2007).

Presurgical Planning

Presurgical planning should include input from all members of the surgical team (e.g., the surgeon, anesthetist, veterinarian, surgical technicians, animal care staff, and investigator). The surgical plan should identify personnel, their roles and training needs, and equipment and supplies required for the procedures planned (Cunliffe-Beamer 1993); the location and nature of the facilities in which the procedures will be conducted; and perioperative animal health assessment and care (Brown and Schofield 1994). A veterinarian should be involved in discussions of the selection of anesthetic agents and doses as well as the plan for perioperative analgesic use. If a nonsterile part of an animal, such as the gastrointestinal tract, is to be surgically exposed or if a procedure is likely to cause immunosuppression, preoperative antibiotics may be appropriate (Klement et al. 1987); however, the routine use of antibiotics should never be considered a replacement for proper aseptic surgical techniques.

Presurgical planning should specify the requirements for postsurgical monitoring, care, and recordkeeping, including the personnel who will perform these duties. The investigator and veterinarian share responsibility for ensuring that postsurgical care is appropriate.

Surgical Facilities

Unless an exception is specifically justified as an essential component of the research protocol and approved by the IACUC, aseptic surgery should be conducted in dedicated facilities or spaces. When determining the appropriate location for a surgical procedure (either a dedicated operating room/suite or an area that provides separation from other activities), the choice may depend on the species, the nature of the procedure (major, minor, or emergency), and the potential for physical impairment or postoperative complications, such as infection. Most bacteria are carried on airborne particles or fomites, so surgical facilities should be maintained and operated in a manner that ensures cleanliness and minimizes unnecessary traffic (AORN 2006; Bartley 1993). If it is necessary to use an operating room for other purposes, it is imperative that the room be returned to an appropriate level of hygiene before its use for major survival surgery.

Generally, agricultural animals maintained for biomedical research should undergo surgery with techniques and in facilities compatible with the guidelines set forth in this section. However, some minor and emergency procedures commonly performed in clinical veterinary practice and in commercial agricultural settings may take place under field conditions. Even when conducted in an agricultural setting, however, these procedures require the use of appropriate aseptic technique, sedatives, analgesics, anesthetics, and conditions commensurate with the risk to the animal's health and well-being.

Surgical Procedures

Surgical procedures are categorized as major or minor and, in the laboratory setting, can be further divided into survival and nonsurvival. As a general guideline, major survival surgery (e.g., laparotomy, thoracotomy, joint replacement, and limb amputation) penetrates and exposes a body cavity, produces substantial impairment of physical or physiologic functions, or involves extensive tissue dissection or transection (Brown et al. 1993). Minor survival surgery does not expose a body cavity and causes little or no physical impairment; this category includes wound suturing, peripheral vessel cannulation, percutaneous biopsy, routine agricultural animal procedures such as castration, and most procedures routinely done on an "outpatient" basis in veterinary clinical practice. Animals recovering from these minor procedures typically do not show significant signs of postoperative pain, have minimal complications, and return to normal function in a relatively short time. When attempting to categorize a particular surgical procedure, the following should be considered: the potential for pain and other postoperative complications; the nature of the procedure as well as the size and location of the incision(s); the duration of the procedure; and the species, health status, and age of the animal.

Laparoscopic surgeries and some procedures associated with neuroscience research (e.g., craniotomy, neurectomy) may be classified as major or minor surgery depending on their impact on the animal (Devitt et al. 2005; Hancock et al. 2005; NRC 2003; Perret-Gentil et al. 1999, 2000). For example, laparoscopic techniques with minimal associated trauma and sequelae (e.g., avian sexing and oocyte collection) could be considered minor, whereas others (e.g., hepatic lobectomy and cholecystectomy) should be considered major. Although minor laparoscopic procedures are often performed on an "outpatient" basis, appropriate aseptic technique, instruments, anesthesia, and analgesia are necessary. Whether a laparoscopic procedure is deemed major or minor should be evaluated on a case-by-case basis by the veterinarian and IACUC.

Emergency situations sometimes require immediate surgical attention under less than ideal conditions. For example, if an animal maintained outdoors needs surgical attention, movement to a surgical facility might be impractical or pose an unacceptable risk to the animal. Such situations often require more intensive aftercare and may pose a greater risk of postoperative complications. The appropriate course of action requires veterinary medical judgment.

In nonsurvival surgery, an animal is euthanized before recovery from anesthesia. It may not be necessary to follow all the techniques outlined in this section if nonsurvival surgery is performed but, at a minimum, the surgical site should be clipped, the surgeon should wear gloves, and the instruments and surrounding area should be clean (Slattum et al. 1991). For nonsurvival procedures of extended duration, attention to aseptic technique may be more important in order to ensure stability of the model and a successful outcome.

Aseptic Technique

Aseptic technique is used to reduce microbial contamination to the lowest possible practical level (Mangram et al. 1999). No procedure, piece of equipment, or germicide alone can achieve that objective (Schonholtz 1976): aseptic technique requires the input and cooperation of everyone who enters the surgery area (Belkin 1992; McWilliams 1976). The contribution and importance of each practice varies with the procedure. Regardless of the species, aseptic technique includes preparation of the patient, such as hair or feather removal and disinfection of the operative site (Hofmann 1979); preparation of the surgeon, such as the provision of appropriate surgical attire, face masks, and sterile surgical gloves (Chamberlain and Houang 1984; Pereira et al. 1990; Schonholtz 1976); sterilization of instruments, supplies, and implanted materials (Bernal et al. 2009; Kagan 1992b); and the use of operative techniques to reduce the likelihood of infection (Ayliffe 1991; Kagan 1992a; Lovaglio and Lawson 1995; Ritter and Marmion 1987; Schofield 1994; Whyte 1988).

While the species of animal may influence the manner in which principles of aseptic technique are achieved (Brown 1994; Cunliffe-Beamer 1983; Gentry and French 1994), inadequate or improper technique may lead to subclinical infections that can cause adverse physiologic and behavioral responses (Beamer 1972; Bradfield et al. 1992; Cunliffe-Beamer 1990; Waynforth 1980, 1987) affecting surgical success, animal well-being, and research results (Cooper et al. 2000). General principles of aseptic technique should be followed for all survival surgical procedures (ACLAM 2001).

Specific sterilization methods should be selected on the basis of the physical characteristics of the materials to be sterilized (Callahan et al. 1995; Schofield 1994) and sterilization indicators should be used to validate that materials have been properly sterilized (Berg 1993). Autoclaving and plasma and gas sterilization are effective methods most commonly used to sterilize instruments and materials. Alternative methods, used primarily for rodent surgery, include liquid chemical sterilants and dry heat sterilization. Liquid chemical sterilants should be used with appropriate contact times and instruments should be rinsed with sterile water or saline before use. Bead or dry heat sterilizers are an effective and convenient means of rapidly sterilizing the working surfaces of surgical instruments but care should be taken to ensure that the instrument surfaces have cooled sufficiently before touching animal tissues to minimize the risk of burns. Alcohol is neither a sterilant nor a high-level disinfectant (Rutala 1990) but may be acceptable for some procedures if prolonged contact times are used (Huerkamp 2002).

Intraoperative Monitoring

Careful monitoring and timely attention to problems increase the likelihood of a successful surgical outcome (Kuhlman 2008). Monitoring includes routine evaluation of anesthetic depth and physiologic functions and conditions, such as body temperature, cardiac and respiratory rates and pattern (Flegal et al. 2009), and blood pressure (Kuhlman 2008), and should be appropriately documented. Use of balanced anesthesia, including the addition of an intraoperative analgesic agent, can help minimize physiologic fluctuations during surgery. Maintenance of normal body temperature minimizes cardiovascular and respiratory disturbances caused by anesthetic agents (Dardai and Heavner 1987; Flegal et al. 2009; Fox et al. 2008), and is of particular importance in small animals where the high ratio of surface area to body weight may easily lead to hypothermia. Fluid replacement may be a necessary component of intraoperative therapy depending on the duration and nature of the procedure. For aquatic species (including amphibians), care should be taken to keep the skin surfaces moist and minimize drying during surgical procedures.

Postoperative Care

An important component of postsurgical care is observation of the animal and intervention as necessary during recovery from anesthesia and surgery (Haskins and Eisele 1997). The intensity of monitoring will vary with the species and the procedure and may be greater during the immediate anesthetic recovery period. During this period, animals should be in a clean, dry, and comfortable area where they can be observed frequently by trained personnel. Particular attention should be given to thermoregulation, cardiovascular and respiratory function, electrolyte and fluid balance, and management of postoperative pain or discomfort. Additional care may be warranted, including long-term administration of parenteral fluids, analgesics, and other drugs, as well as care of surgical incisions. Appropriate medical records should also be maintained.

After recovery from anesthesia, monitoring is often less intense but should include attention to basic biologic functions of intake and elimination and to behavioral signs of postoperative pain, monitoring for postsurgical infections, monitoring of the surgical incision site for dehiscence, bandaging as appropriate, and timely removal of skin sutures, clips, or staples (UFAW 1989).

PAIN AND DISTRESS

An integral component of veterinary medical care is prevention or alleviation of pain associated with procedural and surgical protocols. Pain is a complex experience that typically results from stimuli that damage or have the potential to damage tissue; such stimuli prompt withdrawal and evasive action. The ability to experience and respond to pain is widespread in the animal kingdom and extends beyond vertebrates (Sherwin 2001).

Pain is a stressor and, if not relieved, can lead to unacceptable levels of stress and distress in animals. Furthermore, unrelieved pain may lead to "wind-up," a phenomenon in which central pain sensitization results in a pain response to otherwise nonpainful stimuli (allodynia; Joshi and Ogunnaike 2005). For these reasons, the proper use of anesthetics and analgesics in research animals is an ethical and scientific imperative. *Recognition and Alleviation of Pain in Laboratory Animals* (NRC 2009a) is an excellent source of information about the basis and control of distress and pain (see also Appendix A, Anesthesia, Pain, and Surgery).

Fundamental to the relief of pain in animals is the ability to recognize its clinical signs in specific species (Bateson 1991; Carstens and Moberg 2000; Hawkins 2002; Holton et al. 1998; Hughes and Lang 1983; Karas et al. 2008; Martini et al. 2000; Roughan and Flecknell 2000, 2003, 2004; Sneddon 2006). Species vary in their response to pain (Baumans et al. 1994; Kohn et al. 2007; Morton et al. 2005; Viñuela-Fernández et al. 2007), and criteria for assessing pain in various species differ. The U.S. Government Principles for the Utilization and Care of Vertebrate Animals Used in Testing, Research, and Training (see Appendix B) state that in general, unless the contrary is known or established, it should be considered that procedures that cause pain in humans may also cause pain in other animals (IRAC 1985).

Certain species-specific behavioral manifestations are used as indicators of pain or distress—for example, vocalization (dogs), depression (all), anorexia (all), rapid or labored respiration (rodents, birds, fish), lack of grooming (mammals and birds), increased aggression (mammals and birds), periocular and nasal porphyrin discharge (rodents), abnormal appearance or posture (all), and immobility (all) (NRC 2008, 2009a). However, some species may mask signs of pain until they are quite severe (NRC 2009a). It is therefore essential that personnel caring for and using animals be trained in species-specific and individual clinical, behavioral, physiologic, and biochemical indicators of well-being (Dubner 1987; Karas 2002; Murrell and Johnson 2006; Rose 2002; Stoskopf 1994; Valverde and Gunkel 2005).

Distress may be defined as an aversive state in which an animal fails to cope or adjust to various stressors with which it is presented. But distress may not induce an immediate and observable pathologic or behavioral alteration, making it difficult to monitor and evaluate the animal's state when it is present. Both the duration and intensity of the state are important considerations when trying to prioritize attention to and treatment of animal distress. For example, an injection requiring brief immobilization may produce acute stress lasting only seconds, while long-term individual housing of a social species in a metabolic cage may produce chronic distress. Implementation of clear, appropriate, and humane experimental endpoints for animals, combined with close observation during invasive periods of experimentation, will assist in minimizing distress experienced by animals used in research, teaching, testing, and production. *Recognition and Alleviation of Distress in Laboratory Animals* (NRC 2008) is a resource with important information about distress in experimental animals.

ANESTHESIA AND ANALGESIA

The selection of appropriate analgesics and anesthetics should reflect professional veterinary judgment as to which best meets clinical and humane requirements as well as the needs of the research protocol. The selection depends on many factors, such as the species, age, and strain or stock of the animal, the type and degree of pain, the likely effects of particular agents on specific organ systems, the nature and length of the surgical or pain-inducing procedure, and the safety of the agent, particularly if a physiologic deficit is induced by a surgical or other experimental procedure (Kona-Boun et al. 2005).

Preemptive analgesia (the administration of preoperative and intraoperative analgesia) enhances intraoperative patient stability and optimizes postoperative care and well-being by reducing postoperative pain (Coderre et al. 1993; Hedenqvist et al. 2000). Analgesia may be achieved through timely enteral or parenteral administration of analgesic agents as well as by blocking nociceptive signaling via local anesthetics (e.g., bupivacaine).

Alleviation of chronic pain may be more challenging than postprocedural pain; commercially available opiate slow-release transdermal patches or implantable analgesic-containing osmotic minipumps may be useful for such relief. Because of wide individual variation in response to analgesics, regardless of the initial plan for pain relief, animals should be closely monitored during and after painful procedures and should receive additional drugs, as needed, to ensure appropriate analgesic management (Karas et al. 2008; Paul-Murphy et al. 2004). Nonpharmacologic control of pain may be effective and should not be overlooked as an element of postprocedural or perioperative care for research animals (NRC 2009a; Spinelli 1990). Appropriate nursing support may include a quiet, darkened recovery or resting place, timely wound or bandage maintenance, increased ambient warmth and a soft resting surface, rehydration with oral or parenteral fluids, and a return to normal feeding through the use of highly palatable foods or treats.

Most anesthetics cause a dose-dependent depression of physiologic homeostasis and the changes can vary considerably with different agents. The level of consciousness, degree of antinociception (lack of response to noxious stimuli), and status of the cardiovascular, respiratory, musculoskeletal, and thermoregulatory systems should all be used to assess the adequacy of the anesthetic regimen. Interpretation and appropriate response to the various parameters measured require training and experience with the anesthetic regimen and the species. Loss of consciousness occurs at a light plane of anesthesia, before antinociception, and is sufficient for purposes of restraint or minor, less invasive procedures, but painful stimuli can induce a return to consciousness. Antinociception occurs at a surgical plane of anesthesia and must be ascertained before surgery. Individual animal responses vary widely and a single physiologic or nociceptive reflex response may not be adequate for assessing the surgical plane or level of analgesia (Mason and Brown 1997).

For anesthesia delivery, precision vaporizers and monitoring equipment (e.g., pulse oximeter for determining arterial blood oxygen saturation levels) increase the safety and choices of anesthetic agents for use in rodents and other small species. For injectable anesthetic protocols, specific reversal agents can minimize the incidence of some side effects related to prolonged recovery and recumbency. Guidelines for the selection and proper use of analgesic and anesthetic drugs should be developed and periodically reviewed and updated as standards and techniques are refined. Agents that provide anesthesia and analgesia must be used before their expiration dates and should be acquired, stored, their use recorded, and disposed of legally and safely.

Some classes of drugs such as sedatives, anxiolytics, and neuromuscular blocking agents may not provide analgesia but may be useful when used in combination with appropriate analgesics and anesthetics to provide balanced anesthesia and to minimize stress associated with perioperative procedures. Neuromuscular blocking agents (e.g., pancuronium) are sometimes used to paralyze skeletal muscles during surgery in which general anesthetics have been administered (Klein 1987); because this paralysis eliminates many signs and reflexes used to assess anesthetic depth, autonomic nervous system changes (e.g., sudden changes in heart rate and blood pressure) can be indicators of pain related to an inadequate depth of anesthesia. It is imperative that any proposed use of neuromuscular blocking drugs be carefully evaluated by the veterinarian and IACUC to ensure the well-being of the animal. Acute stress is believed to be a consequence of paralysis in a conscious state and it is known that humans, if conscious, can experience distress when paralyzed with these drugs (NRC 2008; Van Sluyters and Oberdorfer 1991). If paralyzing agents are to be used, the appropriate amount of anesthetic should first be defined on the basis of results of a similar procedure using the anesthetic without a blocking agent (NRC 2003, 2008, 2009a).

EUTHANASIA

Euthanasia is the act of humanely killing animals by methods that induce rapid unconsciousness and death without pain or distress. Unless a deviation is justified for scientific or medical reasons, methods should be consistent with the *AVMA Guidelines on Euthanasia* (AVMA 2007 or later editions). In evaluating the appropriateness of methods, some of the criteria that should be considered are ability to induce loss of consciousness and death with no or only momentary pain, distress, or anxiety; reliability; irreversibility; time required to induce unconsciousness; appropriateness for the species and age of the animal; compatibility with research objectives; and the safety of and emotional effect on personnel.

Euthanasia may be planned and necessary at the end of a protocol or as a means to relieve pain or distress that cannot be alleviated by analgesics, sedatives, or other treatments. Criteria for euthanasia include protocol-specific endpoints (such as degree of a physical or behavioral deficit or tumor size) that will enable a prompt decision by the veterinarian and the investigator to ensure that the endpoint is humane and, whenever possible, the scientific objective of the protocol is achieved (see Chapter 2).

Standardized methods of euthanasia that are predictable and controllable should be developed and approved by the AV and IACUC. Euthanasia should be carried out in a manner that avoids animal distress. Automated systems for controlled and staged delivery of inhalants may offer advantages for species killed frequently or in large numbers, such as rodents (McIntyre et al. 2007). Special consideration should be given to euthanasia of fetuses and larval life forms depending on species and gestational age (Artwohl et al. 2006).

The selection of specific agents and methods for euthanasia will depend on the species involved, the animal's age, and the objectives of the protocol. Generally, chemical agents (e.g., barbiturates, nonexplosive inhalant anesthetics) are preferable to physical methods (e.g., cervical dislocation, decapitation, use of a penetrating captive bolt); however, scientific considerations may preclude the use of chemical agents for some protocols.

Although carbon dioxide (CO_2) is a commonly used method for rodent euthanasia, there is ongoing controversy about its aversive characteristics as an inhalant euthanasia agent. This is an area of active research (Conlee et al. 2005; Danneman et al. 1997; Hackbarth et al. 2000; Kirkden et al. 2008; Leach et al. 2002; Niel et al. 2008) and further study is needed to optimize the methods for CO_2 euthanasia in rodents (Hawkins et al. 2006). The acceptability of CO_2 as a euthanasia agent for small rodents should be evaluated as new data become available. Furthermore, because neonatal rodents are resistant to the hypoxia-inducing effects of CO_2 and require longer exposure times to the agent (Artwohl et al. 2006), alternative methods should be considered (e.g., injection with chemical agents, cervical dislocation, or decapitation; Klaunberg et al. 2004; Pritchett-Corning 2009).

It is essential that euthanasia be performed by personnel skilled in methods for the species in question and in a professional and compassionate manner. Special attention is required to ensure proficiency when a physical method of euthanasia is used. Death must be confirmed by personnel trained to recognize cessation of vital signs in the species being euthanized. A secondary method of euthanasia (e.g., thoracotomy or exsanguination) can be also used to ensure death. All methods of euthanasia should be reviewed and approved by the veterinarian and IACUC.

Euthanizing animals is psychologically difficult for some animal care, veterinary, and research personnel, particularly if they perform euthanasia repetitively or are emotionally attached to the animals being euthanized (Arluke 1990; NRC 2008; Rollin 1986; Wolfle 1985). When delegating euthanasia responsibilities, supervisors should be sensitive to this issue.

REFERENCES

Anderson LC. 2007. Institutional and IACUC responsibilities for animal care and use education and training programs. ILAR J 48:90-95.

ACLAM [American College of Laboratory Animal Medicine]. 2001. Position Statement on Rodent Surgery. Available at www.aclam.org/education/guidelines/position_rodentsurgery.html; accessed January 7, 2010.

AORN [Association of Operating Room Nurses]. 2006. Recommended practices for traffic patterns in the perioperative practice setting. AORN J 83:681-686.

Arluke A. 1990. Uneasiness among laboratory technicians. Lab Anim 19:20-39.

Arndt SS, Lohavech D, van't Klooster J, Ohl F. 2010. Co-species housing in mice and rats: Effects on physiological and behavioural stress responsivity. Horm Behav 57:342-351.
Artwohl J, Brown P, Corning B, Stein S (ACLAM Task Force). 2006. Report of the ACLAM Task Force on Rodent Euthanasia. JAALAS 45:98-105.
ASR [Academy of Surgical Research]. 2009. Guidelines for training in surgical research with animals. J Invest Surg 22:218-225.
AVMA [American Veterinary Medical Association]. 2002. A Report from the American Veterinary Medical Association Animal Air Transportation Study Group. Schaumburg, IL: AVMA.
AVMA. 2007. AVMA Guidelines on Euthanasia. Schaumburg, IL: AVMA.
Ayliffe GAJ. 1991. Role of the environment of the operating suite in surgical wound infection. Rev Infect Dis 13(Suppl 10):S800-S804.
Balaban RS, Hampshire VA. 2001. Challenges in small animal noninvasive imaging. ILAR J 42:248-262.
Barahona H, Melendez LV, Hunt RD, Forbes M, Fraser CEO, Daniel MD. 1975. Experimental horizontal transmission of *herpesvirus saimiri* from squirrel monkeys to an owl monkey. J Infect Dis 132:694-697.
Bartley JM. 1993. Environmental control: Operating room air quality. Today's OR Nurse 15:11-18.
Bateson P. 1991. Assessment of pain in animals. Anim Behav 42:827-839.
Baumans V, Brain PF, Brugere H, Clausing P, Jeneskog T, Perretta G. 1994. Pain and distress in laboratory rodents and lagomorphs. Lab Anim 28:97-112.
Beamer TC. 1972. Pathological changes associated with ovarian transplantation. In: 44th Annual Report of the Jackson Laboratory, Bar Harbor, Maine.
Belkin NJ. 1992. Barrier materials, their influence on surgical wound infections. AORN J 55:1521-1528.
Berg J. 1993. Sterilization. In: Slatter D, ed. Textbook of Small Animal Surgery, 2nd ed. Philadelphia: WB Saunders. p 124-129.
Bernal J, Baldwin M, Gleason T, Kuhlman S, Moore G, Talcott M. 2009. Guidelines for rodent survival surgery. J Invest Surg 22:445-451.
Besselsen DG, Franklin CL, Livingston RS, Riley LK. 2008. Lurking in the shadows: Emerging rodent infectious diseases. ILAR J 49:277-290.
Bradfield JF, Schachtman TR, McLaughlin RM, Steffen EK. 1992. Behavioral and physiological effects of inapparent wound infection in rats. Lab Anim Sci 42:572-578.
Brown MJ. 1994. Aseptic surgery for rodents. In: Niemi SM, Venable JS, Guttman HN, eds. Rodents and Rabbits: Current Research Issues. Bethesda, MD: Scientists Center for Animal Welfare. p 67-72.
Brown MJ, Schofield JC. 1994. Perioperative care. In: Bennett BT, Brown MJ, Schofield JC, eds. Essentials for Animal Research: A Primer for Research Personnel. Washington: National Agricultural Library. p 79-88.
Brown MJ, Pearson PT, Tomson FN. 1993. Guidelines for animal surgery in research and teaching. Am J Vet Res 54:1544-1559.
Butler TM, Brown BG, Dysko RC, Ford EW, Hoskins DE, Klein HJ, Levin JL, Murray KA, Rosenberg DP, Southers JL, Swenson RB. 1995. Medical management. In: Bennett BT, Abee CR, Hendrickson R, eds. Nonhuman Primates in Biomedical Research: Biology and Management. San Diego: Academic Press. p 255-334.
Callahan BM, Hutchinson KA, Armstrong AL, Keller LSF. 1995. A comparison of four methods for sterilizing surgical instruments for rodent surgery. Contemp Top Lab Anim Sci 34:57-60.
Capitanio JP, Kyes RC, Fairbanks LA. 2006. Considerations in the selection and conditioning of Old World monkeys for laboratory research: Animals from domestic sources. ILAR J 47:294-306.

Carstens E, Moberg GP. 2000. Recognizing pain and distress in laboratory animals. ILAR J 41:62-71.

Chamberlain GV, Houang E. 1984. Trial of the use of masks in gynecological operating theatre. Ann R Coll Surg Engl 66:432-433.

Clifford CB, Watson J. 2008. Old enemies, still with us after all these years. ILAR J 49:291-302.

Coderre TJ, Katz J, Vaccarino AL, Melzack R. 1993. Contribution of central neuroplasticity to pathological pain: Review of clinical and experimental evidence. Pain 52:259-285.

Conlee KM, Stephens ML, Rowan AN, King LA. 2005. Carbon dioxide for euthanasia: Concerns regarding pain and distress, with special reference to mice and rats. Lab Anim (NY) 39:137-161.

Conour LA, Murray KA, Brown MJ. 2006. Preparation of animals for research: Issues to consider for rodents and rabbits. ILAR J 47:283-293.

Cooper DM, McIver R, Bianco R. 2000. The thin blue line: A review and discussion of aseptic technique and postprocedural infections in rodents. Contemp Top Lab Anim Sci 39:27-32.

Cunliffe-Beamer TL. 1983. Biomethodology and surgical techniques. In: Foster HL, Small JD, Fox JG, eds. The Mouse in Biomedical Research, vol III: Normative Biology, Immunology and Husbandry. New York: Academic Press. p 419-420.

Cunliffe-Beamer TL. 1990. Surgical techniques. In: Guttman HN, ed. Guidelines for the Well-Being of Rodents in Research. Bethesda, MD: Scientists Center for Animal Welfare. p 80-85.

Cunliffe-Beamer TL. 1993. Applying principles of aseptic surgery to rodents. AWIC Newsletter 4:3-6.

Danneman PJ, Stein S, Walshaw SO. 1997. Humane and practical implications of using carbon dioxide mixed with oxygen for anesthesia or euthanasia of rats. Lab Anim Sci 47:376-385.

Dardai E, Heavner JE. 1987. Respiratory and cardiovascular effects of halothane, isoflurane and enflurane delivered via a Jackson-Rees breathing system in temperature controlled and uncontrolled rats. Meth Find Exptl Clin Pharmacol 9:717-720.

Devitt CM, Cox RE, Hailey JJ. 2005. Duration, complications, stress, and pain of open ovariohysterectomy versus a simple method of laparoscopic-assisted ovariohysterectomy in dogs. JAVMA 227:921-927.

DOI [Department of the Interior]. 2007. Revision of Regulations Implementing the Convention on International Trade in Endangered Species of Wild Fauna and Flora (CITES) (50 CFR Parts 10, 13, 17, and 23). Available at www.fws.gov/policy/library/2007/07-3960.pdf; accessed April 8, 2010.

Dubner R. 1987. Research on pain mechanisms in animals. JAVMA 191:1273-1276.

FASS [Federation of Animal Science Societies]. 2010. Transport. In: Guide for the Care and Use of Agricultural Animals in Research and Teaching, 3rd ed. Champlain, IL: FASS. p. 54.

Field K, Bailey M, Foresman LL, Harris RL, Motzel SL, Rockar RA, Ruble G, Suckow MA. 2007. Medical records for animals used in research, teaching and testing: Public statement from the American College of Laboratory Animal Medicine. ILAR J 48:37-41.

Flegal MC, Fox LK, Kuhlman SM. 2009. Principles of anesthesia monitoring: Respiration. J Invest Surg 22:452-454.

Fox LK, Flegal MC, Kuhlman SM. 2008. Principles of anesthesia monitoring: Body temperature. J Invest Surg 21:373-374.

Gentry SJ, French ED. 1994. The use of aseptic surgery on rodents used in research. Contemp Top Lab Anim Sci 33:61-63.

Hackbarth H, Kuppers N, Bohnet W. 2000. Euthanasia of rats with carbon dioxide: Animal welfare aspects. Lab Anim 34:91-96.

Haines DC, Gorelick PL, Battles JK, Pike KM, Anderson RJ, Fox JG, Taylor NS, Shen Z, Dewhirst FE, Anver MR, Ward JM. 1998. Inflammatory large bowel disease in immunodeficient rats naturally and experimentally infected with *Helicobacter bilis*. Vet Pathol 35:202-208.

Hancock RB, Lanz OI, Waldron DR, Duncan RB, Broadstone RV, Hendrix PK. 2005. Comparison of postoperative pain after ovariohysterectomy by harmonic scalpel-assisted laparoscopy compared with median celiotomy and ligation in dogs. Vet Surg 34:273-282.

Haskins SC, Eisele PH. 1997. Postoperative support and intensive care. In: Kohn DF, Wixson SK, White WJ, Benson GJ, eds. Anesthesia and Analgesia in Laboratory Animals. New York: Academic Press. p 381-382.

Hawkins P. 2002. Recognizing and assessing pain, suffering and distress in laboratory animals: A survey of current practice in the UK with recommendations. Lab Anim 36:378-395.

Hawkins P, Playle L, Golledge H, Leach M, Banzett R, Coenen A, Cooper J, Danneman P, Flecknell P, Kirkden R, Niel L, Raj M. 2006. Newcastle Consensus Report on Carbon Dioxide Euthanasia of Laboratory Animals. Available at www.nc3rs.org.uk/downloaddoc.asp?id=416andpage=292andskin=0; accessed April 10, 2010.

Hedenqvist P, Roughan JV, Flecknell PA. 2000. Effects of repeated anaesthesia with ketamine/medetomidine and of pre-anaesthetic administration of buprenorphine in rats. Lab Anim 34:207-211.

Heon H, Rousseau N, Montgomery J, Beauregard G, Choiniere M. 2006. Establishment of an operating room committee and a training program to improve aseptic techniques for rodent and large animal surgery. JAALAS 45:58-62.

Hirsch VM, Zack PM, Vogel AP, Johnson PR. 1991. Simian immunodeficiency virus infection of macaques: End-stage disease is characterized by wide-spread distribution of proviral DNA in tissues. J Infect Dis 163:976-988.

Hofmann LS. 1979. Preoperative and operative patient management. In: Wingfield WE, Rawlings CA, eds. Small Animal Surgery: An Atlas of Operative Technique. Philadelphia: WB Saunders. p 14-22.

Holton LL, Scott EM, Nolan AM, Reid J, Welsh E, Flaherty D. 1998. Comparison of three methods used for assessment of pain in dogs. JAVMA 212:61-66.

Huerkamp MJR. 2002. Alcohol as a disinfectant for aseptic surgery of rodents: Crossing the thin blue line? Contemp Top Lab Anim Sci 41:10-12.

Hughes HC, Lang CM. 1983. Control of pain in dogs and cats. In: Kitchell RL, Erickson HH, eds. Animal Pain: Perception and Alleviation. Bethesda, MD: American Physiological Society. p 207-216.

Hunt RD, Melendez LV. 1966. Spontaneous herpes-T infection in the owl monkey (*Aotus trivirgatus*). Pathol Vet 3:1-26.

IATA [International Air Transport Association]. 2009. Live Animal Regulations (LAR), 36th ed. Available at www.iata.org/ps/publications/Pages/live-animals.aspx; accessed May 15, 2010.

IRAC [Interagency Research Animal Committee]. 1985. U.S. Government Principles for Utilization and Care of Vertebrate Animals Used in Testing, Research, and Training. Federal Register, May 20. Washington: Office of Science and Technology Policy.

Jacoby RO, Lindsey R. 1998. Risks of infections among laboratory rats and mice at major biomedical research institutions. ILAR J 39:266-271.

Joshi G, Ogunnaike B. 2005. Consequences of inadequate postoperative pain relief and chronic persistent postoperative pain. Anesthesiol Clin North America 23:21-36.

Kagan KG. 1992a. Aseptic technique. Vet Tech 13:205-210.

Kagan KG. 1992b. Care and sterilization of surgical equipment. Vet Tech 13:65-70.

Kagira JM, Ngotho M, Thuita JK, Maina NW, Hau J. 2007. Hematological changes in vervet monkeys (*Chlorocebus aethiops*) during eight months' adaptation to captivity. Am J Primatol 69:1053-1063.

Karas AZ. 2002. Postoperative analgesia in the laboratory mouse, *Mus musculus*. Lab Anim (NY) 31:49-52.

Karas A, Danneman P, Cadillac J. 2008. Strategies for assessing and minimizing pain. In: Fish R, Brown M, Danneman P, Karas A, eds. Anesthesia and Analgesia in Laboratory Animals. San Diego: Academic Press. p 195-218.

Kirkden RD, Niel L, Stewart SA, Weary DM. 2008. Gas killing of rats: The effect of supplemental oxygen on aversion to carbon dioxide. Anim Welf 17:79-87.

Klaunberg BA, O'Malley J, Clark T, Davis JA. 2004. Euthanasia of mouse fetuses and neonates. Contemp Top Lab Anim Sci 43:29-34.

Klein L. 1987. Neuromuscular blocking agents. In: Short CE, ed. Principles and Practice of Veterinary Anesthesia. Baltimore: Williams and Wilkins. p 134-153.

Klement P, del Nido PJ, Mickleborough L, MacKay C, Klement G, Wilson GJ. 1987. Techniques and postoperative management for successful cardiopulmonary bypass and open-heart surgery in dogs. JAVMA 190:869-874.

Kohn DF, Martin TE, Foley PL, Morris TH, Swindle MM, Vogler GA, Wixson SK. 2007. Public statement: Guidelines for the assessment and management of pain in rodents and rabbits. JAALAS 46:97-108.

Kona-Boun JJ, Silim A, Troncy E. 2005. Immunologic aspects of veterinary anesthesia and analgesia. JAVMA 226:355-363.

Kuhlman SM. 2008. Principles of anesthesia monitoring: Introduction. J Invest Surg 21:161-162.

Landi MS, Kreider JW, Lang CM, Bullock LP. 1982. Effects of shipping on the immune function in mice. Am J Vet Res 43:1654-1657.

Leach MC, Bowell VA, Allan TF, Morton DB. 2002. Aversion to gaseous euthanasia agents in rats and mice. Comp Med 52:249-257.

Lerche NW, Yee JL, Capuano SV, Flynn JL. 2008. New approaches to tuberculosis surveillance in nonhuman primates. ILAR J 49:170-178.

Lovaglio J, Lawson PT. 1995. A draping method for improving aseptic technique in rodent surgery. Lab Anim (NY) 24:28-31.

Maggio-Price L, Shows D, Waggie K, Burich A, Zeng W, Escobar S, Morrissey P, Viney JL. 2002. *Helicobacter bilis* infection accelerates and *H. hepaticus* infection delays the development of colitis in multiple drug resistance-deficient (*mdr1a–/–*) mice. Am J Pathol 160:739-751.

Maher JA, Schub T. 2004. Laboratory rodent transportation supplies. Lab Animal 33(8):29-32.

Mangram AJ, Horan ML, Pearson L, Silver C, Jarvis WR. 1999. Guidelines for prevention of surgical site infection, 1999. Infect Control Hosp Epidemiol 20:247-278.

Martini L, Lorenzini RN, Cinotti S, Fini M, Giavaresi G, Giardino R. 2000. Evaluation of pain and stress levels of animals used in experimental research. J Surg Res 88:114-119.

Mason DE, Brown MJ. 1997. Monitoring of anesthesia. In: Kohn DF, Wixson SK, White WJ, Benson GJ, eds. Anesthesia and Analgesia in Laboratory Animals. San Diego: Academic Press.

McIntyre AR, Drummond RA, Riedel ER, Lipman NS. 2007. Automated mouse euthanasia in an individually ventilated caging system: System development and assessment. JAALAS 46:65-73.

McWilliams RM. 1976. Divided responsibilities for operating room asepsis: The dilemma of technology. Med Instrum 10:300-301.

Morton CM, Reid J, Scott EM, Holton LL, Nolan AM. 2005. Application of a scaling model to establish and validate an interval level pain scale for assessment of acute pain in dogs. Am J Vet Res 66:2154-2166.

Murphey-Corb M, Martin LN, Rangan SRS, Baskin GB, Gormus BJ, Wolf RH, Andes WA, West M, Montelaro RC. 1986. Isolation of an HTLV-III-related retrovirus from macaques with simian AIDS and its possible origin in asymptomatic mangabeys. Nature 321:435-437.

Murphy BL, Maynard JE, Krushak DH, Fields RM. 1971. Occurrence of a carrier state for *Herpesvirus tamarinus* in marmosets. Appl Microbiol 21:50-52.

Murrell JC, Johnson CB. 2006. Neurophysiological techniques to assess pain in animals. J Vet Pharmacol Ther 29:325-335.

Nicklas W, Kraft V, Meyer B. 1993. Contamination of transplantable tumors, cell lines, and monoclonal antibodies with rodent viruses. Lab Anim Sci 43:296-299.

Niel L, Stewart SA, Weary DM. 2008. Effect of flow rate on aversion to gradual fill carbon dioxide exposure in rats. Appl Animl Behav Sci 109:77-84.

NRC [National Research Council]. 1991a. Barrier programs. In: Infectious Diseases of Mice and Rats. Washington: National Academy Press. p 17-20.

NRC. 1991b. Individual disease agents and their effects on research. In: Infectious Diseases of Mice and Rats. Washington: National Academy Press. p 31-256.

NRC. 1991c. Health surveillance programs. In: Infectious Diseases of Mice and Rats. Washington: National Academy Press. p 21-27.

NRC. 1996. Rodents: Laboratory Animal Management. Washington: National Academy Press.

NRC. 2003. Guidelines for the Care and Use of Mammals in Neuroscience and Behavioral Research. Washington: National Academies Press.

NRC. 2006. Guidelines for the Humane Transportation of Research Animals. Washington: National Academies Press.

NRC. 2008. Recognition and Alleviation of Distress in Laboratory Animals. Washington: National Academies Press.

NRC. 2009a. Recognition and Alleviation of Pain in Laboratory Animals. Washington: National Academies Press.

NRC. 2009b. Scientific and Humane Issues in the Use of Random Source Dogs and Cats in Research. Washington: National Academies Press.

Obernier JA, Baldwin RL. 2006. Establishing an appropriate period of acclimatization following transportation of laboratory animals. ILAR J 47:364-369.

Otto G, Tolwani RJ. 2002. Use of microisolator caging in a risk-based mouse import and quarantine program: A retrospective study. Contemp Top Lab Anim Sci 41:20-27.

Paul-Murphy J, Ludders JW, Robertson SA, Gaynor JS, Hellyer PW, Wong PL. 2004. The need for a cross-species approach to the study of pain in animals. JAVMA 224:692-697.

Pereira LJ, Lee GM, Wade KJ. 1990. The effect of surgical handwashing routines on the microbial counts of operating room nurses. Am J Infect Control 18:354-364.

Perret-Gentil M, Sinanan M, Dennis MB Jr, Horgan S, Weyhrich J, Anderson D, Hudda K. 1999. Videoendoscopy: An effective and efficient way to perform multiple visceral biopsies in small animals. J Invest Surg 12:157-165.

Perret-Gentil MI, Sinanan MN, Dennis MB Jr, Anderson DM, Pasieka HB, Weyhrich JT, Birkebak TA. 2000. Videoendoscopic techniques for collection of multiple serial intra-abdominal biopsy specimens in HIV-negative and HIV-positive pigtail macaques (*Macaca nemestrina*). J Invest Surg 13:181-195.

Peterson NC. 2008. From bench to cageside: Risk assessment for rodent pathogen contamination of cells and biologics. ILAR J 49:310-315.

Prasad SB, Gatmaitan R, O'Connell RC. 1978. Effect of a conditioning method on general safety test in guinea pigs. Lab Anim Sci 28:591-593.

Pritchett-Corning KR. 2009. Euthanasia of neonatal rats with carbon dioxide. JAALAS 48:23-27.

Pritchett-Corning KR, Chang FT, Festing MF. 2009. Breeding and housing laboratory rats and mice in the same room does not affect the growth or reproduction of either species. JAALAS 48:492-498.

Renquist D. 1990. Outbreak of simian hemorrhagic fever. J Med Primatol 19:77-79.

Ritter MA, Marmion P. 1987. The exogenous sources and controls of microorganisms in the operating room. Orthop Nurs 7:23-28.

Roberts RA, Andrews K. 2008. Nonhuman primate quarantine: Its evolution and practice. ILAR J 49:145-156.

Robertshaw D. 2004. Temperature regulation and the thermal environment. In: Duke's Physiology of Domestic Animals, 12th ed. WO Reese, ed. Ithaca, NY: Cornell University Press.

Robinson V, Anderson S, Carver JFA, Francis RJ, Hubrecht R, Jenkins E, Mathers KE, Raymond R, Rosewell I, Wallace J, Wells DJ. 2003. Refinement and reduction in production of genetically modified mice. Sixth report of the BVAAWF/FRAME/RSPCA/UFAW Joint Working Group on Refinement. Lab Anim 37(Suppl 1):S1-S51. Available at www.lal.org.uk/pdffiles/Transgenic.pdf; accessed January 7, 2010.

Rollin B. 1986. Euthanasia and moral stress. In: DeBellis R, ed. Loss, Grief and Care. Binghamton NY: Haworth Press.

Rose JD. 2002. The neurobehavioral nature of fishes and the question of awareness and pain. Rev Fish Sci 10:1-38.

Roughan JV, Flecknell PA. 2000. Behavioural effects of laparotomy and analgesic effects of ketoprofen and carprofen in rats. Pain 90:65-74.

Roughan JV, Flecknell PA. 2003. Evaluation of a short duration behaviour-based post-operative pain scoring system in rats. Eur J Pain 7:397-406.

Roughan JV, Flecknell PA. 2004. Behaviour-based assessment of the duration of laparotomy-induced abdominal pain and the analgesic effects of carprofen and buprenorphine in rats. Behav Pharmacol 15:461-472.

Rutala WA. 1990. APIC guideline for selection and use of disinfectants. Am J Infect Contr 18:99-117.

Sanhouri AA, Jones RS, Dobson H. 1989. The effects of different types of transportation on plasma cortisol and testosterone concentrations in male goats. Br Vet J 145:446-450.

Schofield JC. 1994. Principles of aseptic technique. In: Bennett BT, Brown MJ, Schofield JC, eds. Essentials for Animal Research: A Primer for Research Personnel. Washington: National Agricultural Library. p 59-77.

Schonholtz GJ. 1976. Maintenance of aseptic barriers in the conventional operating room. J Bone Joint Surg 58:439-445.

Schrama JW, van der Hel W, Gorssen J, Henken AM, Verstegen MW, Noordhuizen JP. 1996. Required thermal thresholds during transport of animals. Vet Q 18(3):90-95.

Sherwin CM. 2001. Can invertebrates suffer? Or, how robust is argument by analogy? Anim Welf 10:103-118.

Slattum MM, Maggio-Price L, DiGiacomo RF, Russell RG. 1991. Infusion-related sepsis in dogs undergoing acute cardiopulmonary surgery. Lab Anim Sci 41:146-150.

Sneddon LU. 2006. Ethics and welfare: Pain perception in fish. Bull Eur Assoc Fish Pathol 26:6-10.

Spinelli J. 1990. Preventing suffering in laboratory animals. In: Rollin B, Kesel M, eds. The Experimental Animal in Biomedical Research, vol I: A Survey of Scientific and Ethical Issues for Investigators. Boca Raton, FL: CRC Press. p 231-242.

Stevens CA, Dey ND. 2007. A program for simulated rodent surgical training. Lab Anim (NY) 36:25-31.

Stoskopf MK. 1994. Pain and analgesia in birds, reptiles, amphibians, and fish. Invest Ophthalmol Vis Sci 35:775-780.

Suckow MA, Doerning BJ. 2007. Assessment of Veterinary Care. In: Silverman J, Suckow MA, Murthy S, eds. The IACUC Handbook, 2nd ed. Boca Raton, FL: CRC Press.

Tuli JS, Smith JA, Morton DB. 1995. Stress measurements in mice after transportation. Lab Anim 29:132-138.

UFAW [Universities Federation for Animal Welfare]. 1989. Surgical procedures. In: Guidelines on the Care of Laboratory Animals and Their Use for Scientific Purposes III. London. p 3-15.

USDA [US Department of Agriculture]. 1985. 9 CFR 1A. (Title 9, Chapter 1, Subchapter A): Animal Welfare. Available at http://ecfr.gpoaccess.gov/cgi/t/text/text-idx?sid=8314313bd7adf2c9f1964e2d82a88d92andc=ecfrandtpl=/ecfrbrowse/Title09/9cfrv1_02.tpl; accessed January 14, 2010.

USDA. 1997. APHIS Policy #3, "Veterinary Care" (July 17, 1997). Available at www.aphis.usda.gov/animal_welfare/downloads/policy/policy3.pdf; accessed January 9, 2010.

Valverde A, Gunkel CI. 2005. Pain management in horses and farm animals. J Vet Emerg Crit Care 15:295-307.

Van Sluyters RC, Oberdorfer MD, eds. 1991. Preparation and Maintenance of Higher Mammals During Neuroscience Experiments. Report of a National Institutes of Health Workshop. NIH No. 91-3207. Bethesda, MD.

Viñuela-Fernández I, Jones E, Welsh EM, Fleetwood-Walker SM. 2007. Pain mechanisms and their implication for the management of pain in farm and companion animals. Vet J 174:227-239.

Waynforth HB. 1980. Experimental and Surgical Technique in the Rat. London: Academic Press.

Waynforth HB. 1987. Standards of surgery for experimental animals. In: Tuffery AA, ed. Laboratory Animals: An Introduction for New Experimenters. Chichester: Wiley-Interscience. p 311-312.

Whyte W. 1988. The role of clothing and drapes in the operating room. J Hosp Infect 11(Suppl C):2-17.

Wolfle TL. 1985. Laboratory animal technicians: Their role in stress reduction and human-companion animal bonding. Vet Clin N Am Small Anim Pract 15:449-454.

5

Physical Plant

GENERAL CONSIDERATIONS

A well-planned, well-designed, well-constructed, properly maintained and managed facility is an important element of humane animal care and use as it facilitates efficient, economical, and safe operation (see Appendix A, Design and Construction of Animal Facilities). The design and size of an animal facility depend on the scope of institutional research activities, the animals to be housed, the physical relationship to the rest of the institution, and the geographic location.

Effective planning and design should include input from personnel experienced with animal facility design, engineering, and operation, as well as from representative users of the proposed facility. Computational fluid dynamics (CFD), building information modeling, and literature on postoccupancy analysis of space use may provide benefits when designing facilities and caging (Eastman et al. 2008; Reynolds 2008; Ross et al. 2009). An animal facility should be designed and constructed in accord with all applicable building codes; in areas with substantial seismic activity the building planning and design should incorporate the recommendations of the Building Seismic Safety Council (BSSC 2001; Vogelweid et al. 2005). Because animal model development and use can be expected to change during the life cycle of an animal facility, facilities should be designed to accommodate changes in use. Modular units (such as custom-designed trailers or prefabricated structures) should comply with construction guidelines described in this chapter.

Building materials for animal facilities should be selected to facilitate efficient and hygienic operation. Durable, moisture- and vermin-proof, fire-resistant, seamless materials are most desirable for interior surfaces, which should be highly resistant to the effects of cleaning agents, scrubbing, high-pressure sprays, and impact. Paints and glazes should be nontoxic if used on surfaces with which animals will have direct contact. In the construction of outdoor facilities, consideration should be given to surfaces that withstand the elements and can be easily maintained.

Location

Quality animal management and human comfort and health protection require separation of animal facilities from personnel areas, such as offices and conference rooms. Separation can be accomplished by having the animal quarters in a separate building, wing, floor, or room. Careful planning should make it possible to place animal housing areas next to or near research laboratories but separated from them by barriers, such as entry locks, corridors, or floors. Additional considerations include the impact of noise and vibration generated from within the facility and from surrounding areas of the building, as well as security of the facility.

Animals should be housed in facilities dedicated to or assigned for that purpose, not in laboratories merely for convenience. If animals must be maintained in a laboratory to satisfy the scientific aims of a protocol, that space should be appropriate to house and care for the animals and its use limited to the period during which it is required. If needed, measures should be taken to minimize occupational hazards related to exposure to animals both in the research area and during transport to and from the area.

Centralization Versus Decentralization

In a physically centralized animal facility, support, care, and use areas are adjacent to the animal housing space. Decentralized animal housing and use occur in space that is not solely dedicated to animal care or support or is physically separated from the support areas and animal care personnel. Centralization often reduces operating costs, providing a more efficient flow of animal care supplies, equipment, and personnel; more efficient use of environmental controls; and less duplication of support services. Centralization reduces the needs for transporting animals between housing and study sites, thereby minimizing the risks of transport stress and exposure to disease agents; affords greater security by providing the opportunity to control facility access; and increases the ease of monitoring staff and animals.

Decentralized animal facilities generally cost more to construct because of the requirement for specialized environmental systems and controls in multiple sites. Duplicate equipment (e.g., cage washers) may be needed, or soiled materials may need to be moved distances for processing. But decentralization may be preferred for certain specialized research services such as imaging, quarantine, and proximity to research facilities, or for biosecurity reasons. Decentralization may be necessary to accommodate large or complex equipment, such as magnetic resonance imaging, or to permit space sharing by users from multiple facilities or institutions. The opportunity for exposure to disease agents is much greater in these situations and special consideration should be given to biosecurity, including transportation to and from the site, quarantine before or after use of the specialized research area, and environmental and equipment decontamination.

The decisions leading to selection of physically centralized versus decentralized animal facilities should be made early and carefully and should involve all stakeholders (NRC 1996; Ruys 1991).

FUNCTIONAL AREAS

Professional judgment should be exercised in the development of a practical, functional, and efficient physical plant for animal care and use. The size, nature, and intensity of an institutional Program (see Chapter 2) will determine the specific facility and support functions needed. In facilities that are small, maintain few animals, or maintain animals under special conditions—such as facilities used exclusively for housing gnotobiotic or specific pathogen-free (SPF) colonies or animals in runs, pens, or outdoor housing—some functional areas listed below may be unnecessary or may be included in a multipurpose area.

Space is required for the following:

- animal housing, care, and sanitation
- receipt, quarantine, separation, and/or rederivation of animals
- separation of species or isolation of individual projects when necessary
- storage.

Most multipurpose animal facilities may also include the following:

- specialized laboratories or space contiguous with or near animal housing areas for such activities as surgery, intensive care, necropsy, irradiation, preparation of special diets, experimental procedures, behavioral testing, imaging, clinical treatment, and diagnostic laboratory procedures

- containment facilities or equipment, if hazardous biologic, physical, or chemical agents are to be used
- barrier facilities for housing of SPF rodents, especially valuable genetically modified animals, or irreplaceable animal models
- receiving and storage areas for food, bedding, pharmaceuticals, biologics, and supplies
- space for washing and sterilizing equipment and supplies and, depending on the volume of work, machines for washing cages, bottles, glassware, racks, and waste cans; a utility sink; a sterilizer for equipment, food, and bedding; and separate areas for holding soiled and clean equipment
- space for storing wastes before incineration or removal
- space for cold storage or disposal of carcasses
- space for administrative and supervisory personnel, including space for staff training and education
- showers, sinks, lockers, toilets, and break areas for personnel
- security features, such as card-key systems, electronic surveillance, and alarms
- areas for maintenance and repair of specialized animal housing systems and equipment.

CONSTRUCTION GUIDELINES

Corridors

Corridors should be wide enough to facilitate the movement of personnel and equipment; a width of 6 to 8 feet can accommodate the needs of most facilities. Floor-wall junctions should be designed to facilitate cleaning. Protective rails or bumpers are recommended and, if provided, should be sealed or manufactured to prevent vermin access. In corridors leading to dog or swine housing facilities, cage-washing facilities, and other high-noise areas, double-door entry vestibules or other noise traps should be considered. Similar entries are advisable for areas leading to nonhuman primate housing as a means to reduce the potential for escape. Double-door entry vestibules also permit air locks in these and other areas where directional airflow is critical for containment or protection. Wherever possible, water lines, drainpipes, reheat coils and valves, electric service connections, and other utilities should be accessible via interstitial space or through access panels or chases in corridors outside the animal rooms. Fire alarms, fire extinguishers, and telephones should be recessed, installed high enough, or shielded by protective guards to prevent damage from the movement of large equipment.

Animal Room Doors

Doors should be large enough (approximately 42 × 84 in.) to allow the easy passage of racks and equipment and they should fit tightly in their frames. Both doors and frames should be appropriately sealed to prevent vermin entry or harborage. Doors should be constructed of and, where appropriate, coated with materials that resist corrosion. Self-closing doors equipped with recessed or shielded handles, sweeps, and kickplates and other protective hardware are usually preferable. Hospital or terminated stops are useful to aid in cleaning (Harris 2005). For safety, doors should open into animal rooms; if it is necessary that they open toward a corridor, there should be a recessed vestibule.

Where room-level security is necessary or it is desirable to limit access (as with the use of hazardous agents), room doors should be equipped with locks or electronic security devices. For personnel safety, doors should be designed to open from the inside without a key.

Doors with viewing windows may be needed for safety and other reasons, but the ability to cover these windows may be considered if exposure to light or hallway activities would be undesirable (e.g., to avoid disturbing the animals' circadian rhythm). Red-tinted windows, which do not transmit specific wavelengths of visible light between corridors and animal rooms, have proved useful for mouse and rat holding rooms as both species have a limited ability to detect light in the red portions of the spectrum (Jacobs et al. 2001; Lyubarsky et al. 1999; Sun et al. 1997).

Exterior Windows

The presence of windows in an animal facility, particularly in animal rooms, creates a potential security risk and should generally be avoided. Windows also create problems with temperature control of the area and prevent strict control of the photoperiod, which is often required in animal-related protocols (and is a critical consideration in rodent breeding colonies). However, in specific situations, windows can provide environmental enrichment for some species, such as nonhuman primates.

Floors

Floors should be moisture resistant, nonabsorbent, impact resistant, and relatively smooth, although textured surfaces may be required in some high-moisture areas and for some species (e.g., farm animals). Floors should be easy to repair and resistant to both the action of urine and other biologic materials and the adverse effects of hot water and cleaning agents. They should be capable of supporting racks, equipment, and stored items without becoming gouged, cracked, or pitted. Depending on their use, floors should be monolithic or have a minimal number of joints. Some materials that have proved satisfactory are epoxy resins, hard-surface sealed concrete, methyl methacrylate, polyurethane, and special hardened rubber-base aggregates. The latter are useful in areas where noise reduction is important. Correct installation is essential to ensure the long-term stability of the surface. If sills are installed at the entrance to a room, they should be designed to allow for convenient passage of equipment.

Drainage

Where floor drains are used, the floors should be sloped and drain traps kept filled with liquid. To minimize prolonged increases in humidity, drainage should allow rapid removal of water and drying of surfaces (Gorton and Besch 1974). Drainpipes should be at least 4 in. (10.2 cm) in diameter, although in some areas, such as dog kennels and agricultural animal facilities, larger drainpipes (≥6 in.) are recommended. A rim- and/or trap-flushing drain or an in-line comminutor may be useful for the disposal of solid waste. When drains are not in use for long periods, they should be capped and sealed to prevent backflow of sewer gases, vermin, and other contaminants; lockable drain covers may be advisable for this purpose in some circumstances.

Floor drains are not essential in all animal rooms, particularly those housing rodents. Floors in such rooms can be sanitized satisfactorily by wet vacuuming or mopping with appropriate cleaning compounds or disinfectants. But the installation of floor drains that are capped when not in use may provide flexibility for future housing of nonrodent species.

Walls and Ceilings

Walls and ceilings should be smooth, moisture resistant, nonabsorbent, and resistant to damage from impact. They should be free of cracks, unsealed utility penetrations, and imperfect junctions with doors, ceilings, floors, walls, and corners. Surface materials should be capable of withstanding cleaning with detergents and disinfectants and the impact of water under high pressure. The use of curbs, guardrails or bumpers, and corner guards should be considered to protect walls and corners from damage, and such items should be solid or sealed to prevent access and harborage of vermin.

Ceilings formed by the concrete slab above are satisfactory if they are smooth and sealed or painted. Suspended ceilings are generally undesirable in animal holding rooms unless they are sealed from the space above with gaskets and clips. When used, they should be fabricated of impervious materials, have a washable surface, and be free of imperfect junctions. Exposed plumbing, ductwork, and light fixtures are undesirable unless the surfaces can be readily cleaned.

Heating, Ventilation, and Air Conditioning (HVAC)

A properly designed and functioning HVAC system is essential to provide environmental and space pressurization control. Temperature and humidity control minimizes variations due either to changing climatic conditions or to differences in the number and kind of animals and equipment in an animal holding space (e.g., a room or cubicle). Pressurization assists in controlling airborne contamination and odors by providing directional airflow between spaces. Areas for quarantine, housing and use of animals exposed to hazardous materials, and housing of nonhuman primates should be kept under relative negative pressure, whereas areas for surgery or clean equipment storage should be kept under relative positive pressure with clean air.

HVAC systems should be designed for reliability (including redundancy where applicable), ease of maintenance, and energy conservation; able to meet requirements for animals as discussed in Chapter 3; and flexible and adaptable to the changing types and numbers of animals and equipment maintained during the life of the facility (ASHRAE 2007a). They should be capable of adjustments in and ideally maintain dry-bulb temperatures of ±1°C (±2°F). Relative humidity should generally be maintained within a range of 30-70% throughout the year. Although maintenance of humidification within a limited range over extended periods is extremely difficult, daily fluctuations (recognizing the effects of routine husbandry especially when caring for large animal species) in relative humidity should be minimized; if excursions outside the desired range are infrequent, minimal, and of short duration, they are unlikely to negatively affect animal well-being. Ideally relative humidity should be maintained within ±10% of set point; however, this may not be achievable under some circumstances.

Constant-volume systems have been most commonly used in animal facilities, but variable-volume (VAV) systems may offer design and operational advantages, such as allowing ventilation rates to be set in accordance with heat load and other variables. These systems offer considerable advantages with respect to flexibility and energy conservation (see Chapter 3).

Previously specified temperature and humidity ranges can be modified to meet special animal needs in circumstances in which all or most of the animal facility is designed exclusively for acclimated species with similar requirements (e.g., when animals are held in a sheltered or outdoor facility). In addition, modifications may need to take into account the microenvironment in some primary enclosures, such as rodent isolator cages, where humidity and temperature may exceed room levels.

Temperature is best regulated by having thermostatic control for each holding space. Use of zonal control for multiple spaces can result in temperature variations between spaces in the zone because of differences in animal densities and heat gain or loss in ventilation ducts and other surfaces within the zone. Individual space control is generally accomplished by providing each space with a dedicated reheat coil. Valves controlling reheat coils should fail in the closed position; steam coils should be avoided or equipped with a high-temperature cut-off system to prevent space overheating and animal loss with valve failure.

Humidification is typically controlled and supplemented on a system or zone basis. Control of humidification in individual holding spaces may be desirable for selected species with reduced tolerance for low relative (e.g., nonhuman primates) or high humidity (e.g., rabbits).

Most HVAC systems are designed for average high and low temperatures and humidities experienced in a geographic area within ±5% variation (ASHRAE 2009). Moderate fluctuations in temperature and relative humidity outside suggested ranges are generally well tolerated by most species commonly used in research as long as they are brief and infrequent; holding spaces should be designed to minimize drafts and temperature gradients. Consideration should be given to measures that minimize fluctuations in temperature and relative humidity outside the recommended ranges due to extremes in the external ambient environment. Such measures can include partial redundancy, partial air recirculation, altered ventilation rates, or the use of auxiliary equipment. In the event of an HVAC system or component failure, systems should at the minimum supply facility needs at a reduced level, address the adverse effects of loss of temperature control, and, where necessary, maintain critical pressurization gradients. It is essential that life-threatening heat accumulation or loss be prevented during mechanical failure. Temporary needs for ventilation of sheltered or outdoor facilities can usually be met with auxiliary equipment.

Air handling system intake locations should avoid entrainment of fumes from vehicles, equipment, and system exhaust. While 100% outside air is typically provided, when recirculated air is used its quality and quantity should be in accord with recommendations in Chapter 3. The type and efficiency of supply and exhaust air treatment should be matched to the quantity and types of contaminants and to the risks they pose. Supply air is usually filtered with 85–95% dust spot efficient filters (ASHRAE 2008). In certain instances, higher efficiency filters (e.g., HEPA) may be beneficial for recirculated supply air and air supplied to or exhausted from specialized areas such as surgical and containment facilities (Kowalski et al. 2002).

Power and Lighting

The electrical system should be safe and provide appropriate lighting, a sufficient number of power outlets, and suitable amperage for specialized equipment. In the event of power failure, an alternative or emergency power supply should be available to maintain critical services (e.g., the HVAC system, ventilated caging systems [Huerkamp et al. 2003], or life support systems for aquatic species) or support functions (e.g., freezers and isolators) in animal rooms, operating suites, and other essential areas. Consideration should be given to outfitting movable equipment for which uninterrupted power is essential (e.g., ventilated racks), with twist-lock plugs to prevent accidental removal from the power supply.

Light fixtures, timers, switches, and outlets should be properly sealed to prevent vermin access. Recessed energy-efficient fluorescent lights are commonly used in animal facilities. Spectral quality of lights may be important for some species when maintained in the laboratory; in these cases full spectrum lamps may be appropriate. A time-controlled lighting system should be used to ensure a uniform diurnal lighting cycle. Override systems should be equipped with an automatic timeout or a warning light to indicate the system is in override mode, and system performance and override functions should be regularly evaluated to ensure proper cycling. Dual-level lighting may be considered when housing species that are sensitive to high light intensity, such as albino rodents; low-intensity lighting is provided during the light phase of the diurnal cycle, and higher-intensity lighting is provided as needed (e.g., when personnel require enhanced visibility). Light bulbs or fixtures should be equipped with protective covers to ensure the safety of the animals and personnel. Moisture-resistant switches and outlets and ground-fault interrupters should be used in areas with high water use, such as cage-washing areas and aquarium-maintenance areas.

Storage Areas

Adequate space should be available for storage of equipment, supplies, food, bedding, and refuse. Corridors are not appropriate storage areas. Storage space can be decreased when delivery of materials and supplies is reliable and frequent; however, it should be ample enough to accommodate storage of essential commodities to ensure the animals' uninterrupted husbandry and care (e.g., if delivery is delayed). Bedding and food should be stored in a separate area free from vermin and protected from the risk of contamination from toxic or hazardous substances. Areas used for food storage should not be subject to elevated temperatures or relative humidity for prolonged periods. Refuse storage areas should be separated from other storage areas. Refrigerated storage, separated from other cold storage, is essential for storage of dead animals and animal tissue waste; this storage area should be kept below 7°C (44.6°F) to reduce putrefaction of wastes and animal carcasses and should be constructed in a manner that facilitates cleaning.

Noise Control

Noise control is an important consideration in an animal facility and should be addressed during the planning stages of new facility design or renovation (see Chapter 3). Noise-producing support functions, such as cage washing, are commonly separated from housing and experimental functions. Masonry walls, due to their density, generally have excellent sound-attenuating properties, but similar sound attenuation can be achieved using many different materials and partition designs. For example, sanitizable sound-attenuating materials bonded to walls or ceilings may be appropriate for noise control in some situations, whereas acoustic materials applied directly to the ceiling or as part of a suspended ceiling in an animal room present problems for sanitation and vermin control and are not recommended. Experience has shown that well-constructed corridor doors, sound-attenuating doors, or double-door entry vestibules can help to control the transmission of sound along corridors. An excellent resource on partition design for sound control is available in *Noise Control in Buildings: A Practical Guide for Architects and Engineers* (Warnock and Quirt 1994).

Attention should be paid to attenuating noise generated by equipment (ASHRAE 2007b). Fire and environmental-monitoring alarm systems and public address systems should be selected and positioned to minimize potential animal disturbance. The location of equipment capable of generating sound at ultrasonic frequencies is important as some species can hear such high frequencies. Selecting equipment for rodent facilities that does not generate noise in the ultrasonic range should be considered.

Vibration Control

Vibration may arise from mechanical equipment, electrical switches, and other building components, or from remote sources (via groundborne transmission). Regarding the latter, special consideration should be given to the building structure type especially if the animal facility will be located over, under, or adjacent to subways, trains, or automobile and truck traffic. Like noise, different species can detect and be affected by vibrations of different frequencies and wavelengths, so attempts should be made to identify all vibration sources and isolate or dampen them with vibration suppression systems (ASHRAE 2007b).

Facilities for Sanitizing Materials

A dedicated central area for sanitizing cages and ancillary equipment should be provided. Mechanical cage-washing equipment is generally needed and should be selected to match the types of caging and equipment used. Consideration should be given to such factors as the following:

- location with respect to animal rooms and waste disposal and storage areas
- ease of access, including doors of sufficient width to facilitate movement of equipment
- sufficient space for staging and maneuvering of equipment
- soiled waste disposal and prewashing activities
- ease of cleaning and disinfection of the area
- traffic flow that separates animals and equipment moving between clean and soiled areas
- air pressurization between partitioned spaces to reduce the potential of cross contamination between soiled and clean equipment
- insulation of walls and ceilings where necessary
- sound attenuation
- utilities, such as hot and cold water, steam, floor drains, and electric power
- ventilation, including installation of vents or canopies and provisions for dissipation of steam and fumes from sanitizing processes
- vibration, especially if animals are housed directly above, below, or adjacent to the washing facility
- personnel safety, by ensuring that safety showers, eyewash stations, and other equipment are provided as required by code; exposed hot water and steam lines are properly insulated; procedures with a propensity to generate aerosols are appropriately contained; and equipment, such as cage/rack washers, and bulk sterilizers, which personnel enter, are equipped with functioning safety devices that prevent staff from becoming trapped inside.

Environmental Monitoring

Monitoring of environmental conditions in animal holding spaces and other environmentally sensitive areas in the facility should be considered. Automated monitoring systems, which notify personnel of excursions in environmental conditions, including temperature and photoperiod, are advisable to prevent animal loss or physiologic changes as a result of system malfunction. The function and accuracy of such systems should be regularly verified.

SPECIAL FACILITIES

Surgery

The design of a surgical facility should accommodate the species to be operated on and the complexity of the procedures to be performed (Hessler 1991; see also Appendix A, Design and Construction of Animal Facilities). The facility, including that used for rodents, by necessity becomes larger and more complex as the number and size of animals or the complexity of procedures increase. For instance, a larger facility may be required to accommodate procedures on agricultural species, large surgical teams, imaging devices, robotic surgical systems, and/or laparoscopic equipment towers. Surgical facilities for agricultural species may additionally require floor drains, special restraint devices, and hydraulic operating tables.

For most survival surgery performed on rodents and other small species such as aquatics and birds, an animal procedure laboratory is recommended; the space should be dedicated to surgery and related activities when used for this purpose, and managed to minimize contamination from other activities conducted in the room at other times. The association of surgical facilities with diagnostic laboratories, imaging facilities, animal housing, staff offices, and so on should be considered in the overall context of the complexity of the surgical program. Centralized surgical facilities are cost-effective in equipment, conservation of space and personnel resources, and reduced transit of animals. They also enable enhanced personnel safety and professional oversight of both facilities and procedures.

For most surgical programs, functional components of aseptic surgery include surgical support, animal preparation, surgeon's scrub, operating room, and postoperative recovery. The areas that support those functions should be designed to minimize traffic flow and separate the related nonsurgical activities from the surgical procedure in the operating room. The separation is best achieved by physical barriers (AORN 1993) but may also be achieved by distance between areas or by the timing of appropriate cleaning and disinfection between activities.

Surgical facilities should be sufficiently separate from other areas to minimize unnecessary traffic and decrease the potential for contamination (Humphreys 1993). The number of personnel and their level of activity have been shown to be directly related to the level of bacterial contamination and the incidence of postoperative wound infection (Fitzgerald 1979). Traffic in the operating room can be reduced by the installation of an observation window, a communication system (such as an intercom system), and judicious location of doors.

Control of contamination and ease of cleaning should be key considerations in the design of a surgical facility. The interior surfaces should be constructed of materials that are monolithic and impervious to moisture. Ventilation systems supplying filtered air at positive pressure can reduce the risk of postoperative infection (Ayscue 1986; Bartley 1993; Schonholtz 1976). Careful location of air supply and exhaust ducts and appropriate room ventilation rates are also recommended to minimize contamination (Ayliffe 1991; Bartley 1993; Holton and Ridgway 1993; Humphreys 1993). To facilitate cleaning, the operating rooms should have as little fixed equipment as possible (Schonholtz 1976; UFAW 1989). Other operating room features to consider include surgical lights to provide adequate illumination (Ayscue 1986); sufficient electric outlets for support equipment; gases to support anesthesia, surgical procedures, and gas-powered equipment; vacuum; and gas-scavenging capability.

The surgical support area should be designed for washing and sterilizing instruments and for storing instruments and supplies. Autoclaves are commonly placed in this area. It is often desirable to have a large sink in the animal preparation area to facilitate cleaning of the animal and the operating facilities. A dressing area should be available for personnel to change into surgical attire; a multipurpose locker room can serve this function. There should be a scrub area for surgeons, equipped with foot, knee, or electric-eye surgical sinks (Knecht et al. 1981). To minimize the potential for contamination of the surgical site by aerosols generated during scrubbing, the scrub area should usually be outside the operating room and animal preparation area.

A postoperative recovery area should provide the physical environment to support the needs of the animal during the period of anesthetic and immediate postsurgical recovery and should be sited to allow adequate observation of the animal during this period. The electric and mechanical requirements of monitoring and support equipment should be considered. The type of caging and support equipment will depend on the species and types of procedures but should be designed to be easily cleaned and to support physiologic functions, such as thermoregulation and respiration. Depending on the circumstances, a postoperative recovery area for farm animals may be modified or nonexistent in some field situations, but precautions should be taken to minimize risk of injury to recovering animals.

Barrier Facilities

Barrier facilities are designed and constructed to exclude the introduction of adventitious infectious agents from areas where animals of a defined health status are housed and used. They may be a portion of a larger facility or a free-standing unit. While once used primarily for rodent production facilities and to maintain immunodeficient rodents, many newer facilities incorporate barrier features for housing specific pathogen-free (SPF) mice and rats, especially valuable genetically engineered animals, and SPF animals of other species.

Barrier facilities typically incorporate airlock or special entries (e.g., air or wet showers) for staff and supplies. Staff generally wear dedicated clothing and footwear, or freshly laundered, sterile, or disposable outer garments such as gowns, head and shoe covers, gloves, and sometimes face masks prior to entry. Consumables, such as feed or bedding, that may harbor infectious agents are autoclaved or are gamma-irradiated by the supplier and surface decontaminated on entry. Drinking water may be autoclaved or subject to specialized treatment (e.g., reverse osmosis filtration) to remove infectious agents. Caging and other materials with which the animals have direct contact may be sterilized after washing before reuse. Strict operational procedures are frequently established to preclude intermingling of clean and soiled supplies and personnel groups, depending on work function. Only animals of defined health status are received into the barrier, and once they leave they are prohibited from reentering without retesting. Personnel entry is restricted and those with access are appropriately trained in procedures that minimize the introduction of contaminants.

Engineering features may include high-level filtration of supply air (e.g., HEPA or 95% efficient filters), pressurization of the barrier with respect to surrounding areas, and directional airflow from clean to potentially contaminated areas. Specialized equipment augmenting the barrier may include isolator cages, individually ventilated cages, and animal changing stations.

Detailed information on barrier design, construction, and operations has been recently published (Hessler 2008; Lipman 2006, 2008).

Imaging

In vivo imaging offers noninvasive methods for evaluating structure and function at the level of the whole animal, tissue, or cell, and allows for the sequential study of temporal events (Chatham and Blackband 2001; Cherry and Gambhir 2001). Imaging devices vary in the technology used to generate an image, body targets imaged, resolution, hazard exposure, and requirements for use. The devices may be self-shielded and require no modifications of the surrounding structure to operate safely, or they may require concrete, solid core masonry, lead-, steel-, or copper-lined walls, or other construction features to operate safely or minimize interference with devices and activities in adjacent areas. Because imaging devices are often expensive to acquire and maintain, and may require specialized support space and highly trained personnel to operate, shared animal imaging resources may be preferable.

Consideration should be given to the location of the imaging resource. Whether located in the animal facility or in a separate location, cross contamination between groups of animals, different animal species, or between animals and humans (if the device is used for both animal and human subjects) is possible because these devices may be difficult to sanitize (Klaunberg and Davis 2008; Lipman 2006). If the imaging resource is located outside the animal facility, appropriate transportation methods and routes should be developed to avoid inappropriate exposure of humans to animals in transit. If possible, animals should not be moved past offices, lunch rooms, or public areas where people are likely to be present.

As imaging may require the subject to be immobile, often for extended time periods during image acquisition, provisions should be made for delivery of anesthetics and carrier gas, the scavenging of waste anesthetic gas, and adequate animal monitoring (Balaban and Hampshire 2001). Remote storage of gas tanks is generally required in facilities where magnetic resonance (MR) scanners are used as the magnetic field requires ferrous materials to be kept a safe distance away from the magnet. Site selection of MR scanners requires special attention because of their weight, the fringe field generated (especially from unshielded magnets), and the impact of ferrous elements of the building structure or its components, especially those that are not static (e.g., elevators), as they may affect field homogeneity. Most MR scanners are superconducting and require the use of cryogens. Because cryogen boil-off can lead to asphyxiation of both personnel and animals, rooms with MR scanners or in which cryogen gases are stored must be equipped with oxygen sensors and a method for increasing room ventilation to exhaust inert gases during cryogen filling (Klaunberg and Davis 2008).

Many imaging devices, especially those designed for small animals, are self-contained and require no special physical plant considerations. Provisions should be made to locate the operating console away from imaging devices that emit ionizing or magnetic radiation. Imaging devices with components that are difficult to sanitize should be covered with a disposable or sanitizable material when not in use.

Whole Body Irradiation

Total body irradiation of small laboratory animals may be accomplished using devices that emit either gamma- or X-rays. Devices are usually self-shielded and, because of the weight of the shielding material, may require special site considerations. Devices with gamma-emitting sources are subject to regulations that require adherence to specific security, monitoring, and personnel clearance requirements (Nuclear Regulatory Commission 2008). The site selected for irradiators should also take into account whether they are to be used for animals and biologics, as well as the source and microbial status of the animals to be irradiated. Locating them in the animal facility may require access for personnel who would normally not require it or may necessitate bringing animals into a facility where they are not normally housed.

Hazardous Agent Containment

The goal of containment is to "reduce or eliminate exposure of laboratory workers, other persons, and the outside environment to potential hazardous agents" (DHHS 2010). This is accomplished by employing appropriate practices and equipment, vaccinating personnel if a vaccine is available, and ensuring the proper design and operation of the physical plant.

Animal facilities used to study biologic agents that are infectious to humans are categorized into different biosafety levels of escalating containment requirements as described in *Biosafety in Microbiological and Biomedical Laboratories* (BMBL; DHHS 2009 or most recent version). Each animal biosafety level (ABSL) reflects a combination of practices, safety equipment, and facilities based on risk of human infection. As described in the 2009 edition of the BMBL, ABSL-1 contains agents not known to cause human infection; ABSL-2 contains agents of moderate risk that cause human disease by ingestion or percutaneous or mucosal exposure; ABSL-3 contains agents that cause serious and potentially lethal infections and have known potential for aerosol transmission; and ABSL-4 contains nonindigenous (exotic) agents that pose high individual risk of life-threatening disease and for which there is no available vaccine or treatment. Facility design, engineering criteria, construction methods and materials, commissioning, and validation become more important with each increasing level. The BMBL should be consulted for specific design and engineering requirements. Considerable care should be taken when selecting the team of professionals responsible for the design, engineering, construction, and commissioning of a containment facility.

Guidelines have also been developed for containing agricultural pathogens (USDA ARS 2002), recombinant DNA molecules (NIH 2002), arthropod vectors (ACME, ASTMH 2003), and hazardous chemicals (NRC 1995). Biologic agents and toxins pose a threat to animal and plant health or public health and safety, and facilities in which they are used must adhere to APHIS, USDA, and CDC Select Agent Regulations (CFR 2005; CDC and DHHS 1996; PL 107-56; PL 107-188;) and/or other applicable federal, state, or local regulations. These regulations stipulate, among other requirements, that the institution registered to use select agents establish and adhere to stringent security measures.

The specific facility features, equipment, and safety practices to be employed will depend, to a considerable extent, on whether a specific hazard is a particulate, volatile, or both. Facility features applicable to all hazards include isolation of the animals and their waste, provision of sealed monolithic room surfaces that do not promote dust accumulation and are easy to sanitize, increased air exchange rates to dilute environmental contamination if it occurs, air pressure differentials to ensure that areas containing hazards have negative pressure with respect to surrounding areas, specialized housing systems, if available, and appropriate safety equipment such as a biologic safety cabinet or chemical hood (CDC and NIH 2007). A number of references are available to provide an overview of the issues related to hazardous material containment (Frazier and Talka 2005; Lehner et al. 2008; Lieberman 1995; NRC 1989, 1995)

Behavioral Studies

When planning a behavioral facility, special attention should be given to all aspects of facility design, construction, equipment, and use that may generate conditions that inappropriately stimulate the senses of the test animals. It is frequently necessary to maintain animals in an environment, especially during periods of testing and observation, with strict control over auditory, visual, tactile, and olfactory stimuli. The facility site, as well as the engineering and construction methods used, should be carefully selected to minimize airborne transmission of noise and groundborne transmission of vibration.

Noise and vibration may arise from the building's structure, its equipment, or from human activities (see section on Noise). The frequencies and intensity of sound, which stimulate auditory responses in the species being investigated, should guide the selection of construction materials, techniques, and equipment to minimize intrusions. For instance, the HVAC system should be designed and components selected to ensure that noise, including ultrasonic frequencies, is not generated; fire alarm annunciators that emit sound at a frequency not audible to rodents should be used; hardware should be provided on doors to enable them to close quietly; nonessential noise-generating equipment should be housed outside the study area; and personnel traffic should be minimized both in animal testing areas and in areas contiguous to them (Heffner and Heffner 2007). Attention should be given to the control of aberrant visual cues, especially in circadian studies. The selection of the type, intensity, and control of lighting will likely differ from other animal facility areas. A variety of specialized housing and testing systems may also need to be accommodated in the facility.

Special construction features may also be desirable. Double-door vestibule entries to the behavioral facility, testing suites, or individual testing rooms may be useful as they can prevent noise, odors, and light from entering the behavioral testing area. Floor coverings that reduce sound transmission should be selected. Testing rooms may require floor drains, water sources, and increased floor loading to support specific behavioral testing apparatus.

Consideration should be given to the types and amount of electronics and other equipment used to ensure that the HVAC system can accommodate the associated heat loads. Airlocks and air pressure differentials between spaces can provide olfactory segregation of species and activities and thus reduce the risk of altered behavioral responses (ASHRAE 2007c).

When possible, testing equipment should be designed in such a way as to allow surface disinfection between studies. Components that cannot be cleaned or disinfected, such as computers and recording equipment, should be located in areas where contact with animals is unlikely and should be covered when not in use (the use of computer keyboard covers may also be beneficial). Provision of sufficient space for storage of behavioral apparatus and equipment should also be considered. As transportation to and from the testing area may alter behavioral responses, consideration should be given to providing housing areas contiguous with those used for testing; if such areas are provided, they should meet the requirements specified in the *Guide*.

Aquatic Species Housing

Many of the construction features described above are applicable to those for aquatic species, but particular consideration should be given to the housing systems used and the methods for maintaining the aquatic environment.

The complexity of the life support system depends on the species housed and the size, type, and number of tanks and animals supported. All systems require a water source, which may require prior treatment (e.g., ultraviolet sterilization and particulate, carbon, and ultrafiltration). Holding areas for aquatic species should be provided with drains of a suitable size and number to accommodate water released during system operation and maintenance or as a result of life support system or tank failure. Drains should not permit passage of animals or hazardous materials into the sanitary system without appropriate treatment.

Materials used for floors, walls, and ceilings should be impervious to water while floors should be slip resistant and able to withstand the loads inherent with large quantities of water. Electrical receptacles or circuits should be ground-fault interrupted to prevent electrocution of personnel and animals. Doors and frames, supply diffusers, exhaust registers, lighting fixtures, HVAC ducts and components (exposed to high levels of moisture or corrosives), and other metallic elements should be made of moisture- and corrosion-resistant materials. Housing systems, life support system components, and plumbing used to distribute water after treatment, including adhesives to connect components, should be constructed of materials that are nontoxic and biologically inert. If the macroenvironmental/room HVAC system is used as the primary method for tempering the aquatic environment, sufficient ventilation should be provided to prevent moisture buildup on room surfaces and maintain suitable temperatures for the species housed.

SECURITY AND ACCESS CONTROL

Recent episodes of domestic terrorism have heightened awareness of the importance of animal facility security, but there are other reasons why security and access control should be provided. Most animals maintained for research are vulnerable to infection with adventitious agents and therefore access to them should be strictly controlled and made available only to personnel who have received appropriate training and have a legitimate need for access. Animals used in studies with hazardous materials require special precautions for personnel before access, and staff entering the animal facility should have completed the institution's occupational health and safety training.

When possible, the animal facility should be located within another structure with its own independent set of security features. Vehicular access should be limited and, when provided, controlled and monitored.

Security and access control are generally provided in zones, starting at the perimeter with areas of highest security located within other zones. Control measures may consist of security personnel, physical barriers, and control devices. The scope of the security system should depend on the size of the facility as well as the nature of the activities conducted within. Increasingly, access control is extended from the facility's perimeter to each animal holding room. Microprocessor-controlled security systems are frequently employed because of the large number of control points and staff requiring access. These systems typically use electronic key or proximity cards and associated readers, which, in addition to controlling access, enable recording of the time, location, and personal identification of each entry. In more sensitive areas, biometric reading devices such as thumb or palm readers or retinal scanners may be more suitable because key cards can be shared. Security may be enhanced with electronic and video surveillance systems. These systems may be monitored by personnel or motion-activated recording devices.

REFERENCES

ACME, ASTMH [American Committee of Medical Entomology, American Society of Tropical Medicine and Hygiene]. 2003. Arthropod Containment Guidelines. Vector-Borne Zoonotic Dis 3:61-98.

AORN [Association of Operating Room Nurses]. 1993. Recommended practices: Traffic patterns in the surgical suite. AORN J 57:730.

ASHRAE [American Society of Heating, Refrigeration, and Air Conditioning Engineers]. 2007a. Chapter 22: Environmental control for animals and plants. In 2007 ASHRAE Handbook: Fundamentals, I-P ed. Atlanta.

ASHRAE. 2007b. Chapter 47: Sound and vibration control. In 2007 ASHRAE Handbook: HVAC Applications. Atlanta.

ASHRAE. 2007c. Chapter 45: Control of gaseous indoor air contaminants. In 2007 ASHRAE Handbook: HVAC Applications. Atlanta.

ASHRAE. 2008. Chapter 28: Air cleaners for particulate contaminants. In 2008 ASHRAE Handbook: HVAC Systems and Equipment. Atlanta.

ASHRAE. 2009. Chapter 14: Climatic design information. In 2009 ASHRAE Handbook: Fundamentals. Atlanta.

Ayliffe GAJ. 1991. Role of the environment of the operating suite in surgical wound infection. Rev Infect Dis 13(Suppl 10):S800-S804.

Ayscue D. 1986. Operating room design: Accommodating lasers. AORN J 41:1278-1285.

Balaban RS, Hampshire VA. 2001. Challenges in small animal noninvasive imaging. ILAR J 42:248-262.

Bartley JM. 1993. Environmental control: Operating room air quality. Today's OR Nurse 15:11-18.

BSSC [Building Seismic Safety Council]. 2001. National Earthquake Hazards Reduction Program Recommended Provisions for Seismic Regulations for New Buildings and Other Structures, 2000 ed, FEMA 369. Washington: FEMA.

CDC and DHHS [Centers for Disease Control and Prevention and Department of Health and Human Services]. 1996. Additional requirements for facilities transferring or receiving select agents. Final Rule, Federal Register 61:55189-55200. October 24.

CDC and NIH [Centers for Disease Control and Prevention and National Institutes of Health]. 2007. Primary Containment for Biohazards: Selection, Installation and Use of Biological Safety Cabinets, 3rd ed. Washington: Government Printing Office. Available at www.cdc.gov/biosafety/publications/bmbl5/BMBL5_appendixA.pdf; accessed July 15, 2010.

CFR [Code of Federal Regulations]. 2005. 7 CFR Part 331 and 9 CFR Part 121, Agricultural Bioterrorism Protection Act of 2002. USDA, APHIS, Possession, use and transfer of biological agents and toxins. 42 CFR Part 1003, Final Rule, Department of Health and Human Services, March.

Chatham JC, Blackband SJ. 2001. Nuclear magnetic resonance spectroscopy and imaging in animal research. ILAR J 42:189-208.

Cherry SR, Gambhir SS. 2001. Use of positron emission tomography in animal research. ILAR J 42:219-232.

DHHS [Department of Health and Human Services]. 2009. Biosafety in Microbiological and Biomedical Laboratories, 5th ed. Chosewood LC, Wilson DE, eds. Washington: Government Printing Office. Available at http://www.cdc.gov/biosafety/publications/bmbl5/index.htm; accessed July 30, 2010.

Eastman C, Teicholz P, Sacks R, Liston K. 2008. BIM Handbook: A Guide to Building Information Modeling for Owners, Managers, Designers, Engineers, and Contractors. Hoboken, NJ: John Wiley and Sons.

Fitzgerald RH. 1979. Microbiologic environment of the conventional operating room. Arch Surg 114:772-775.

Frazier D, Talka J. 2005. Facility design considerations for select agent animal research. ILAR J 46:23-33.

Gorton RL, Besch EL. 1974. Air temperature and humidity response to cleaning water loads in laboratory animal storage facilities. ASHRAE Trans 80:37-52.

Harris CM. 2005. Dictionary of Architecture and Construction, 4th ed. Columbus, OH: McGraw-Hill.

Heffner HE, Heffner RS. 2007. Hearing ranges of laboratory animals. JAALAS 46:20-22.
Hessler JR. 1991. Facilities to support research. In: Ruys T, ed. Handbook of Facility Planning, vol 2: Laboratory Animal Facilities. New York: Van Nostrand. p 34-55.
Hessler JR. 2008. Barrier housing for rodents. In: Hessler J, Lehner N, eds. Planning and Designing Animal Research Facilities. Amsterdam: Academic Press. p 335-345.
Holton J, Ridgway GL. 1993. Commissioning operating theatres. J Hosp Infect 23:153-160.
Huerkamp MJ, Thompson WD, Lehner NDM. 2003. Failed air supply to individually ventilated caging system causes acute hypoxia and mortality of rats. Contemp Top Lab Anim Sci 42:44-45.
Humphreys H. 1993. Infection control and the design of a new operating theatre suite. J Hosp Infect 23:61-70.
Jacobs GH, Fenwick JA, Williams GA. 2001. Cone-based vision of rats for ultraviolet and visible lights. J Exp Biol 204:2439-2446.
Klaunberg BA, Davis JA. 2008. Considerations for laboratory animal imaging center design and setup. ILAR J 49:4-16.
Knecht CD, Allen AR, Williams DJ, Johnson JH. 1981. Fundamental Techniques in Veterinary Surgery, 2nd ed. Philadelphia: WB Saunders.
Kowalski WJ, Bahnfleth WP, Carey DD. 2002. Engineering control of airborne disease transmission in animal laboratories. Contemp Top Lab Anim Sci 41:9-17.
Lehner NDM, Crane JT, Mottet MD, Fitzgerald ME. 2008. Biohazards: Safety practices, operations, and containment facilities. In: Hessler J, Lehner N, eds. Planning and Designing Research Animal Facilities. London: Academic Press. p 347-364.
Lieberman D. 1995. Biohazards Management Handbook, 2nd ed. New York: Marcel Dekker.
Lipman NS. 2006. Design and management of research facilities for mice. In: Fox J, Barthold S, Newcomer C, Smith A, Quimby F, Davisson M, eds. The Mouse in Biomedical Research, 2nd ed, vol III. London: Academic Press. p 270-317.
Lipman NS. 2008. Rodent facilities and caging systems. In: Hessler J, Lehner N, eds. Planning and Designing Animal Research Facilities. Amsterdam: Academic Press. p 265-288.
Lyubarsky AL, Falsini B, Pennesi ME, Valentini P, Pugh ENJ. 1999. UV- and midwave sensitive cone-driven retinal responses of the mouse: A possible phenotype for coexpression of cone photopigments. J Neurosci 19:442-455.
NIH [National Institutes of Health]. 2002. Guidelines for Research Involving Recombinant DNA Molecules. Available at http://oba.od.nih.gov/rdna/nih_guidelines_oba.html; accessed May 15, 2010.
NRC [National Research Council]. 1989. Biosafety in the Laboratory: Prudent Practices for Handling and Disposal of Infectious Materials. Washington: National Academy Press.
NRC. 1995. Prudent Practices in the Laboratory: Handling and Disposal of Chemicals. Washington: National Academy Press.
NRC. 1996. Laboratory Animal Management: Rodents. Washington: National Academy Press.
Nuclear Regulatory Commission. 2008. NRC Regulatory Issue Summary 2008-02, Actions to Increase the Security of High Activity Radioactive Sources. Washington: NRC Office of Federal and State Materials and Environmental Management Programs.
PL [Public Law] 107-56. 2001. Uniting and Strengthening America by Providing Appropriate Tools Required to Intercept and Obstruct Terrorism (USA PATRIOT) Act of 2001. Washington: Government Printing Office.
PL 107-188. 2002. Public Health Security and Bioterrorism Preparedness and Response Act of 2002. Washington: Government Printing Office.
Reynolds SD. 2008. Using computational fluid dynamics (CFD) in laboratory animal facilities. In: Hessler J, Lehner N, eds. Planning and Designing Research Animal Facilities. Amsterdam: Academic Press. p 479-488.

Ross S, Schapiro S, Hau J, Lukas K. 2009. Space use as an indicator of enclosure appropriateness: A novel measure of captive animal welfare. Appl Anim Behav Sci 121:42-50.
Ruys T. 1991. Handbook of Facilities Planning, vol 2. New York: Van Nostrand Reinhold.
Schonholtz GJ. 1976. Maintenance of aseptic barriers in the conventional operating room. J Bone Joint Surg 58:439-445.
Sun H, Macke JP, Nathans J. 1997. Mechanisms of spectral tuning in the mouse green cone pigment. Proc Natl Acad Sci USA 94:8860-8865.
UFAW [Universities Federation for Animal Welfare]. 1989. Guidelines on the Care of Laboratory Animals and Their Use for Scientific Purposes, III: Surgical Procedures. Herts, UK.
USDA ARS [United States Department of Agriculture Animal Research Services]. 2002. ARS Facilities Design Standards. Pub. 242.01 Facilities Division, Facilities Engineering Branch, AFM/ARS. Washington: Government Printing Office.
Vogelweid CM, Hill JB, Shea RA, Johnson DB. 2005. Earthquakes and building design: A primer for the laboratory animal professional. Lab Anim (NY) 34:35-42.
Warnock ACC, Quirt JD. 1994. Chapter 5: Airborne Sound Insulation and Appendix 5 Tables on Sound Transmission Loss. In: Harris CM, ed. Noise Control in Buildings: A Practical Guide for Architects and Engineers. Columbus OH: McGraw-Hill. p 5.1-5.32; 5.33-5.77.

Addendum

Guide for the Care and Use of Laboratory Animals Eighth Edition

ADDENDUM: LIST OF EDITORIAL CHANGES FROM THE PREPUBLICATION VERSION[1]

1. Page 22. Original sentence: "Personnel training should include information on laboratory animal allergies, preventive control measures and proper techniques for working with animals (Gordon et at. 1997; Schweitzer et al. 2003; Thulin et al. 2002)." "...early recognition and reporting of allergy symptoms" was added to clarify the guidance and reflect the cited references.

2. Page 31. Original sentence: "They should therefore be used, when available, for all animal-related procedures (NIH 2008; USDA 1997b)." The NIH reference was removed because it applies only to the NIH intramural animal research program.

3. Page 32. Original sentence: "Principal investigators conducting field research should be knowledgeable of relevant zoonotic diseases, associated safety issues, and, when working in an international environment, any local laws or regulations that apply." Compliance with laws and regulations applies to field investigations irrespective of location, so "when working in an international environment" and "local" were deleted.

---

[1]Page numbers reflect placement of revisions in this report.

4. Page 32. Original sentence: "Appropriate veterinary input is needed for projects involving capture, individual identification, sedation, anesthesia, surgery, recovery, holding, transportation, release, or euthanasia." Beginning of sentence changed to "Veterinary input may be needed..." to clarify the intent of the sentence.

5. Page 43. Original sentence: "The ambient temperature range in which thermoregulation occurs without the need to increase metabolic heat production is called the *thermoneutral zone (TNZ)* and is bounded by the upper (UCT) and lower critical temperatures (LCT)." The phrase "or activate evaporative heat loss mechanisms" and reference to Gordon 2005 were added to provide a more complete definition of thermoneutral zone.

6. Page 45. Original sentence: "In climates where it is difficult to provide a sufficient level of environmental relative humidity, animals should be closely monitored for negative effects such as excessively flaky skin in birds and mammals, ecdysis (molting) difficulties in reptiles, and desiccation stress in semiaquatic amphibians." Because the significance of excessive flaky skin varies among species, "in birds and mammals" was deleted.

7. Page 55. Original sentence: "Thus there is no ideal formula for calculating an animal's space needs based only on body size or weight." The phrase "and readers should take the performance indices discussed in this section into consideration when utilizing the species-specific guidelines presented in the following pages" was added to further clarify the intent of this section.

8. Pages 57-63. The phrase "the interpretation of this table should take into consideration the performance indices described in the text beginning on page 55" was added as a footnote to tables as additional guidance to their interpretation. Additionally, the symbol "≥" was restored (due to mistaken deletion) in Table 3.2 (mice in groups; rats in groups; hamsters; guinea pigs) and added to Tables 3.5 (group 8) and 3.6 (sheep and goats)

9. Page 69. Original sentence: "Cedar shavings are not recommended because they emit aromatic hydrocarbons that induce hepatic microsomal enzymes and cytotoxicity (Torronen et al. 1989; Weichbrod et al. 1986, 1988)." Text mistakenly deleted was restored to the end of the sentence: "and have been reported to increase the incidence of cancer (Jacobs and Dieter 1978; Vlahakis 1977)."

10. Page 107. Original sentence: "The Centers for Disease Control and Prevention enforces regulations to prevent the introduction, transmission, or spread of communicable diseases and regulate the importation of any animal or animal product capable of carrying a zoonotic disease." Because USDA also has jurisdiction over imports, it was added to the sentence.

11. Page 107. Original sentence: "The US Fish and Wildlife Service regulates importation/exportation and interstate trade of wild vertebrate and invertebrate animals and their tissues." Because US Fish and Wildlife Services does not regulate interstate trade except for species listed under the Endangered Species Act, "and interstate trade" was deleted

12. Page 120. Original sentence: "Additional care might be warranted, including long-term administration of parenteral fluids (FBR 1987), analgesics and other drugs; and care of surgical incisions." Because this is now standard veterinary medical practice and the cited reference is no longer in print, the reference was removed.

13. Page 120. Original sentence: "In general, unless the contrary is known or established, it should be considered that procedures that cause pain in humans may also cause pain in vertebrate species (IRAC 1985)." For consistency with the text of the U.S. Government Principles the phrase "The U.S. Government Principles for the Utilization and Care of Vertebrate Animals Used in Testing, Research, and Training (see Appendix B) state that..." was added and the words "in vertebrate species" were replaced with "in other animals."

14. Page 143. Original phrase: "Vibration, especially if animals are housed directly above the washing facility," was expanded because animals housed below or adjacent to the washing facility are also subject to potential vibration.

15. Added references: pages 163; 169; 179; 182; 192.

Appendices

APPENDIX A

Additional Selected References

SUBJECT MATTER

Nonhuman Primates
Rodents and Rabbits
Other Animals

VETERINARY CARE
Transportation
Anesthesia, Pain, and Surgery
Disease Surveillance, Diagnosis, and Treatment
Pathology, Clinical Pathology, and Parasitology
Species-Specific References—Veterinary Care
Agricultural Animals
Amphibians, Reptiles, and Fish
Birds
Cats and Dogs
Exotic, Wild, and Zoo Animals
Nonhuman Primates
Rodents and Rabbits

DESIGN AND CONSTRUCTION OF ANIMAL FACILITIES

USE OF LABORATORY ANIMALS

Alternatives

Alternative Methods for Toxicity Testing: Regulatory Policy Issues. EPA 230/12 85 029. NTIS PB8 6 113404/AS. Washington: Office of Policy, Planning, and Evaluation, US Environmental Protection Agency.
Alternatives to Animal Use in Research, Testing, and Education. 1986. Office of Technology Assessment (OTA BA 273). Washington: Government Printing Office.
Alternatives to Current Uses of Animals in Research, Safety Testing, and Education. 1986. Stephens ML. Washington: Humane Society of the United States.
Alternatives to Pain in Experiments on Animals. 1980. Pratt D. New York: Argus Archives.
Alternative Toxicological Methods. 2003. Salem H, Katz S. Boca Raton FL: CRC Press.
Animals and Alternatives in Testing: History, Science, and Ethics. 1994. Zurlo J, Rudacille D, Goldberg AM. New York: Mary Ann Liebert Publishers.
Future improvements: Replacement in vitro methods. 2002. Balls M. ILAR J 43(Suppl): S69-S73.
ICCVAM Recommendations on In Vitro Methods for Assessing Acute Systemic Toxicity. 2001. Available at http://iccvam.niehs.nih.gov/docs/acutetox_docs/finalrpt//finappi2.pdf; accessed January 24, 2010.
Regulatory Testing and Animal Welfare. 2002. ILAR J 43(Supplement).
Implementation of the 3Rs (refinement, reduction, and replacement): Validation and regulatory acceptance considerations for alternative toxicological test methods. 2002. Schechtman L. ILAR J 43:S85-S94.
Incorporating the 3Rs into regulatory scientific practices. 2002. Sterling S, Rispin A. ILAR J 43:S18-S20.

Refinement, reduction, and replacement of animal use for regulatory testing: Future improvements and implementation within the regulatory framework. 2002. Richmond J. ILAR J 43:S63-S68.

The Role of the Interagency Coordinating Committee on the Validation of Alternative Methods (ICCVAM) in the Evaluation of New Toxicological Testing Methods. 2000. Stokes WS, Hill RN. In: Proceedings of the 3rd World Congress on Alternatives and Animal Use in the Life Sciences, Bologna, Italy, 1999. New York: Elsevier.

Ethics and Welfare

An additional "R": Remembering the animals. 2002. Iliff SA. ILAR J 43:38-47.

Animal Liberation, 2nd ed. 1990. Singer P. New York: New York Review Book (distributed by Random House).

Animal Rights and Human Obligations, 2nd ed. 1989. Regan T, Singer P. Englewood Cliffs NJ: Prentice-Hall.

Animal Welfare: Competing Conceptions and Their Ethical Implications. 2008. Haynes RP. New York: Springer.

Animals and Why They Matter. 1983. Midgley M. Athens: University of Georgia Press.

Applied Ethics in Animal Research: Philosophy, Regulation, and Laboratory Applications. 2002. Gluck JP, DiPasquale T, Orlans FB. West Lafayette IN: Purdue University Press.

Bioethics in Laboratory Animal Research. 1999. ILAR J 40(1).

Bioethics, animal research, and ethical theory. 1999. Russow L-M. ILAR J 40:15-21.

How and why animals matter. 1999. Donnelley S. ILAR J 40:22-28.

Roots of concern with nonhuman animals in biomedical ethics. 1999. Sideris L, McCarthy CR, Smith DH. ILAR J 40:3-14.

Challenges in assessing fish welfare. 2009. Volpato GL. ILAR J 50:329-337.

Cost of Caring: Recognizing Human Emotions in the Care of Laboratory Animals. 2001. Memphis: American Association for Laboratory Animal Science.

Ethical aspects of relationships between humans and research animals. 2002. Herzog H. ILAR J 43:27-32.

Ethical implications of the human-animal bond in the laboratory. 2002. Russow L-M. ILAR J 43:33-37.

Ethical scores for animal procedures. 1992. Porter D. Nature 356:101-102.

Ethics and pain research in animals. 1999. Tannenbaum J. ILAR J 40:97-110.

Fish and welfare: Do fish have the capacity for pain perception and suffering? 2004. Braithwaite VA, Huntingford FA. Anim Welf 13:S87-S92.

Guidance notes on retrospective review: A discussion document prepared by the LASA Ethics and Training Group. 2004. Jennings M, Howard B. Tamworth UK: Laboratory Animal Science Association.

Guidelines for the ethical use of animals in applied ethology studies. 2003. Sherwin CM, Christiansen SB, Duncan IJ, Erhard HW, Lay DC Jr, Mench JA, O'Connor CE, Petherick JC. Appl Anim Behav Sci 81:291-305.

Guidelines to Promote the Well-being of Animals Used for Scientific Purposes: The Assessment and Alleviation of Pain and Distress in Research Animals. 2008. National Health and Medical Research Council, Australian Government. Available at www.nhmrc.gov.au/_files_nhmrc/file/publications/synopses/ea18.pdf; accessed January 24, 2010.

In the Name of Science: Issues in Responsible Animal Experimentation. 1993. Orlans FB. New York: Oxford University Press.

International Guiding Principles for Biomedical Research Involving Animals. 1985. Council for International Organizations of Medical Sciences (CIOMS). Available at http://www.cioms.ch/publications/guidelines/1985_texts_of_guidelines.htm; accessed October 2, 2010.

Moral Status: Obligations to Persons and Other Living Things. 1997. Warren MA. Gloucestershire: Clarendon Press.

Of Mice, Models, and Men: A Critical Evaluation of Animal Research. 1984. Rowan AN. Albany: State University of New York Press.

Painful dilemmas: The ethics of animal-based pain research. 2009. Magalhaes-Sant'Ana M, Sandoe P, Olsson IAS. Anim Welf 18:49-63

Principles and guidelines for the development of a science-based decision making process facilitating the implementation of the 3Rs by governmental regulators. 2002. Gauthier C. ILAR J 43:S99-S104.

Principles and practice in ethical review of animal experiments across Europe: Summary of the report of the FELASA Working Group on Ethical Evaluation of Animal Experiments. 2007. Smith JA, van den Broek FAR, Canto Martorell J, Hackbarth H, Ruksenas O, Zeller W. Lab Anim 41:143-160.

Recognition and Alleviation of Distress in Laboratory Animals. 2008. National Research Council. Washington: National Academies Press.

Refinement of the use of non-human primates in scientific research, part I: The influence of humans. 2006. Rennie AE, Buchanan-Smith HM. Anim Welf 15:203-213.

Review of Cost-benefit Assessment in the Use of Animals in Research. 2003. Animal Procedures Committee. London. Available at http://apc.homeoffice.gov.uk/reference/costbenefit.pdf; accessed January 24, 2010.

Science, Medicine, and Animals. 2004. National Research Council. Washington: National Academies Press.

Taking Animals Seriously: Mental Life and Moral Status. 1996. DeGrazia D. New York: Cambridge University Press.

The Assessment and "Weighing" of Costs. 1991. In: Smith JA, Boyd K, eds. Lives in the Balance: The Ethics of Using Animals in Biomedical Research. London: Oxford University Press.

The Ethics of Research Involving Animals. 2005. London: Nuffield Council on Bioethics.

The Experimental Animal in Biomedical Research, vol I: A Survey of Scientific and Ethical Issues for Investigators. 1990. Rollin BE, Kesel ML, eds. Boca Raton FL: CRC Press.

The Frankenstein Syndrome: Ethical and Social Issues in the Genetic Engineering of Animals. 1995. Rollin BE. New York: Cambridge University Press.

The regulation of animal research and the emergence of animal ethics: A conceptual history. 2006. Rollin BE. Theor Med Bioeth 27:285-304.

The Three Rs: A journey or a destination? 2000. Richmond J. ATLA 28:761-773.

Experimental Design and Statistics

Animal welfare and the statistical consultant. 1993. Engeman RM, Shumake SA. Am Statistician 47:229-233.

Appropriate animal numbers in biomedical research in light of animal welfare considerations. 1991. Mann MD, Crouse DA, Prentice ED. Lab Anim Sci 41:6-14.

Common errors in the statistical analysis of experimental data. 2002. Festing MFW. In: Balls M, van Zeller A-M, Halder ME, eds. Progress in the Reduction, Refinement and Replacement of Animal Experimentation: Developments in Animal and Veterinary Science. Amsterdam: Elsevier. p 753-758.

Experimental Design and Statistics in Biomedical Research. 2002. ILAR J 43(4).

- Control of variability. 2002. Howard BR. ILAR J 43:194-201.
- Guidelines for the design and statistical analysis of experiments using laboratory animals. 2002. Festing MFW, Altman DG. ILAR J 43:244-258.
- Practical aspects to experimental design in animal research. 2002. Johnson PD, Besselsen DG. ILAR J 43:202-206.

Role of ancillary variables in the design, analysis, and interpretation of animal experiments. 2002. Gaines Das R. ILAR J 43:214-222.
Sample size determination. 2002. Dell RB, Holleran S, Ramakrishnan R. ILAR J 43:207-213.
Use of factorial designs to optimize animal experiments and reduce animal use. 2002. Shaw R, Festing MFW, Peers I, Furlong L. ILAR J 43:223-232.
Primer of Biostatistics, 6th ed. 2005. Glantz SA. New York: McGraw-Hill.
Sample size determination (Appendix A). 2003. National Research Council. In: Guidelines for the Care and Use of Mammals in Neuroscience and Behavioral Research. Washington: National Academies Press. p 175-180.
Statistical Methods, 8th ed. 1989. Snedecor GW, Cochran WG. Ames: Iowa State Press.
The Design and Analysis of Long-Term Animal Experiments. 1986. Gart JJ, Krewski D, Lee PN, Tarone RE, Wahrendorf J. Lyon: International Agency for Research on Cancer.
The Design of Animal Experiments: Reducing the Use of Animals in Research Through Better Experimental Design. 2002. Festing MFW, Overend P, Gaines Das R, Cortina Borja M, Berdoy M. London: Royal Society of Medicine Press.
What is it like to be a rat? Rat sensory perception and its implications for experimental design and rat welfare. 2008. Burn CC. Appl Anim Behav Sci 112:1-32.

Research and Testing Methodology

Adjuvants and Antibody Production. 1995. ILAR J 37(3).
Advanced Physiological Monitoring in Rodents. 2002. ILAR J 43(3).
Mechanical ventilation for imaging the small animal. Hedlund LW, Johnson GA. 2002. ILAR J 43:159-174.
Miniaturization: An overview of biotechnologies for monitoring the physiology and pathophysiology of rodent animal models. 2002. Goode TL, Klein HJ. ILAR J 43:136-146.
Casarett and Doull's Toxicology: The Basic Science of Poisons, 8th ed. 2007. Klaassen CD. New York: McGraw-Hill.
Categorising the Severity of Scientific Procedures on Animals: Summary and Reports from Three Roundtable Discussions. 2004. Smith JA, Jennings M, eds. West Sussex UK: RSPCA, Research Animals Department. Available at www.boyd-group.demon.co.uk/severity_report.pdf; accessed January 24, 2010.
Clinical considerations in rodent bioimaging. 2004. Colby LA, Morenko BJ. Comp Med 54:623-630.
Effects of Freund's complete adjuvant on the physiology, histology, and activity of New Zealand white rabbits. 2004. Halliday LC, Artwohl JE, Bunte RM, Ramakrishnan V, Bennett BT. Contemp Top Lab Anim Sci 43:8-13.
Ethological research techniques and methods. 1998. Novak MA, West M, Bayne KL, Suomi SJ. In: Hart L, ed. Responsible Conduct of Research in Animal Behavior. New York: Oxford University Press. p 51-66.
Genetic Engineering and Animal Welfare: Preparing for the 21st Century. 1999. Gonder JC, Prentice ED, Russow L-M, eds. Greenbelt MD: Scientists Center for Animal Welfare.
Guidance for Industry and Other Stakeholders: Toxicological Principles for the Safety Assessment of Food Ingredients, Redbook 2000 (rev 2007). FDA. Available at www.fda.gov/Food/GuidanceComplianceRegulatoryInformation/GuidanceDocuments/FoodIngredientsandPackaging/Redbook/default.htm; accessed May 15, 2010.
Humane Endpoints for Animals Used in Biomedical Research and Testing. 2000. ILAR J 41(2).

Humane endpoints for laboratory animals used in toxicity testing. 2000. Stokes WS. In: Proceedings of the 3rd World Congress on Alternatives and Animal Use in the Life Sciences, Bologna, Italy, 1999. New York: Elsevier.

Humane endpoints in animal experiments for biomedical research: Proceedings of the International Conference, 22-25 November 1998, Zeist, the Netherlands.

Immunization Procedures and Adjuvant Products. 2005. ILAR J 46(3).

- Adjuvants and antibody production: Dispelling the myths associated with Freund's complete and other adjuvants. 2005. Stills JF Jr. ILAR J 46:280-293.
- Advances in monoclonal antibody technology: Genetic engineering of mice, cells, and immunoglobulins. 2005. Peterson NC. ILAR J 46:314-319.
- Applications and optimization of immunization procedures. 2005. Schunk MK, Macallum GE. ILAR J 46:241-257.
- Monoclonal versus polyclonal antibodies: Distinguishing characteristics, applications, and information. 2005. Lipman NS, Jackson LR, Trudel LJ, Weis-Garcia F. ILAR J 46:258-268.
- Using polyclonal and monoclonal antibodies in regulatory testing of biological products. 2005. Clough NE, Hauer PJ. ILAR J 46:300-306.

Impact of Noninvasive Technology on Animal Research. 2001. ILAR J 42(3).

- Challenges in small animal noninvasive imaging. 2001. Balaban RS, Hampshire VA. ILAR J 42:248-262.
- Nuclear magnetic resonance spectroscopy and imaging in animal research. 2001. Chatham JC, Blackband SJ. ILAR J 42:189-208.
- Use of positron emission tomography in animal research. 2001. Cherry SR, Gambhir SS. ILAR J 42:219-232.

Integration of safety pharmacology endpoints into toxicology studies. 2002. Luft J, Bode G. Fundam Clin Pharmacol 16:91-103.

Joint Working Group on Refinement: Refinements in telemetry procedures. Seventh report of the BVAAWF/FRAME/RSPCA/UFAW Joint Working Group on Refinement, Part A. 2003. Morton DB, Hawkins P, Beyan R, Heath K, Kirkwood J, Pearce P, Scott L, Whelan G, Webb A. Lab Anim 37:261-299. Available at www.rspca.org.uk/ImageLocator/LocateAsset?asset=document&assetId=1232712323332&mode=prd; accessed August 10, 2010.

Methods and Welfare Considerations in Behavioral Research with Animals. 2002. Report of a National Institutes of Health Workshop. NIMH. Available at www.nimh.nih.gov/researchfunding/grants/animals.pdf; accessed January 24, 2010.

Monoclonal Antibody Production. 1999. National Research Council. Washington: National Academy Press.

Physiologic and behavioral assessment of rabbits immunized with Freund's complete adjuvant. 2000. Halliday LC, Artwohl JE, Hanly WC, Bunte RM, Bennett BT. Contemp Top Lab Anim Sci 39:8-13.

Physiological monitoring of small animals during magnetic resonance imaging. 2005. Mirsattari SM, Bihari F, Leung LS, Menon RS, Wang Z, Ives JR, Bartha R. J Neurosci Meth 144:207-213.

Principles and Methods in Toxicology, 4th ed. 2007. Hayes AW. Philadelphia: Taylor and Francis.

Refinement of neuroscience procedures using nonhuman primates. 2005. Wolfensohn S, Peters A. Anim Technol Welf 4:49-50.

Regulatory Testing and Animal Welfare. 2002. ILAR J 43(Suppl).

- Animal care best practices for regulatory testing. 2002. Fillman-Holliday D, Landi MS. ILAR J 43:S49-S58.
- Animal use in the safety evaluation of chemicals: Harmonization and emerging needs. 2002. Spielmann H. ILAR J 43:S11-S17.

Future improvements and implementation of animal care practices within the animal testing regulatory environment. 2002. Guittin P, Decelle T. ILAR J 43:S80-S84.
Incorporating the 3Rs into regulatory scientific practices. 2002. Sterling S, Rispin A. ILAR J 43:S18-S20.
Possibilities for refinement and reduction: Future improvements within regulatory testing. 2002. Stephens ML, Conlee K, Alvino G, Rowan A. ILAR J 43:S74-S79.
Preclinical safety evaluation using nonrodent species: An industry/welfare project to minimize dog use. 2002. Smith D, Broadhead C, Descotes G, Fosse R, Hack R, Krauser K, Pfister R, Phillips B, Rabemampianina Y, Sanders J, Sparrow S, Stephan-Gueldnew M, Jacobsen SD. ILAR J 43:S39-S42.
Refinement, reduction, and replacement of animal use for regulatory testing: Future improvements and implementation within the regulatory framework. 2002. Richmond J. ILAR J 43:S63-S68.
The International Symposium on Regulatory Testing and Animal Welfare: Recommendations on best scientific practices for acute local skin and eye toxicity testing. 2002. Botham PA, Hayes AW, Moir D. ILAR J 43:S105-S107.
The International Symposium on Regulatory Testing and Animal Welfare: Recommendations on best scientific practices for acute systemic toxicity testing. 2002. Stitzel K, Spielmann H, Griffin G. ILAR J 43:S108-S111.
The International Symposium on Regulatory Testing and Animal Welfare: Recommendations on best scientific practices for animal care in regulatory toxicology. 2002. Morris T, Goulet S, Morton D. ILAR J 43:S123-S125.
The International Symposium on Regulatory Testing and Animal Welfare: Recommendations on best scientific practices for biologicals: Safety and potency evaluations. 2002. Cussler K, Kulpa J, Calver J. ILAR J 43:S126-S128.
The International Symposium on Regulatory Testing and Animal Welfare: Recommendations on best scientific practices for subchronic/chronic toxicity and carcinogenicity testing. 2002. Combes R, Schechtman L, Stokes WS, Blakey D. ILAR J 43:S112-S117.
The safety assessment process: Setting the scene—An FDA perspective. 2002. Schechtman L. ILAR J 43:S5-S10.
Tiered testing strategies: Acute local toxicity. 2002. Stitzel K. ILAR J 43:S21-S26.
The use of laboratory animals in toxicologic research. 2001. White WJ. In: Hays AW, ed. Principles and Methods in Toxicology. Philadelphia: Taylor and Francis. p 773-818.
The use of radiotelemetry in small laboratory animals: Recent advances. 2001. Kramer K, Kinter L, Brockway BP, Voss HP, Remie R, VanZutphen BL. Contemp Top Lab Anim Sci 40:8-16.

PROGRAM MANAGEMENT

General References

Cost Analysis and Rate Setting Manual for Animal Resource Facilities. 2000. National Center for Research Resources. Available at www.ncrr.nih.gov/publications/comparative_medicine/CARS.pdf; accessed January 24, 2010.
O Disaster Planning and Management. 2010. ILAR J 51(2).
Crisis planning to manage risks posed by animal rights extremists. 2010. Bailey MR, Rich BA, Bennett BT. ILAR J 51:138-148.
Disaster preparedness in biocontainment animal research facilities: Developing and implementing an incident response plan (IRP). 2010. Swearengen JR, Vargas KJ, Tate MK, Linde NS. ILAR J 51:120-126.

Introduction: Disaster planning and management: A practicum. 2010. Bayne KA. ILAR J 51:101-103.
IACUC considerations: You have a disaster plan but are you really prepared? 2010. Wingfield WE, Rollin BE, Bowen RA. ILAR J 51:164-170.
Management of rodent viral disease outbreaks: One institution's (r)evolution. 2010. Smith AL. ILAR J 51:127-137.
Tropical storm and hurricane recovery and preparedness strategies. 2010. Goodwin BS Jr, Donaho JC. ILAR J 51:104-119.
Verification of poultry carcass composting research through application during actual avian influenza outbreaks. 2010. Flory GA, Peer RW. ILAR J 51:149-157.
Wildfire evacuation: Outrunning the witch's curse—One animal center's experience. 2010. Arms MM, Van Zante JD. ILAR J 51:158-163.
Essentials for Animal Research: A Primer for Research Personnel. 1994. Bennett BT, Brown MJ, Schofield JC. Beltsville MD: National Agricultural Library.
Infectious Disease Research in the Age of Biodefense. ILAR J 46(1).
Administrative issues related to infectious disease research in the age of bioterrorism. 2005. Jaax J. ILAR J 46:8-14.
Public response to infectious disease research: The UC Davis experience. 2005. Fell AH, Bailey PJ. ILAR J 46:66-72.
Laboratory security and emergency response guidance for laboratories working with select agents. Richmond JY, Nesby-O'Dell SL. 2002. MMWR Recomm Rep 51:1-8.
Management of Laboratory Animal Care and Use Programs. 2002. Suckow MA, Douglas FA, Weichbrod RH, eds. Boca Raton FL: CRC Press.
Use of Laboratory Animals in Biomedical and Behavioral Research. 1988. National Research Council and Institute of Medicine. Washington: National Academy Press.
Using site assessment and risk analysis to plan and build disaster-resistant programs and facilities. 2003. Vogelweid CM, Hill JB, Shea RA, Truby SJ, Schantz LD. Lab Anim 32:40-44.

Laws, Regulations, and Policies

Animal Care and Use: Policy Issues in the 1990s. National Institutes of Health/Office for Protection from Research Risks (NIH/OPRR). 1989. Proceedings of NIH/OPRR Conference, Bethesda MD.
Animal Care Policy Manual. APHIS [Animal and Plant Health Inspection Service], USDA. Available at www.aphis.usda.gov/animal_welfare/policy.shtml; accessed May 15, 2010.
Animal Care Resource Guide: Research Facility Inspection Guide. APHIS, USDA. Available at www.aphis.usda.gov/animal_welfare/rig.shtml; accessed May 15, 2010.
Animal Law Section. National Association for Biomedical Research. Available at http://www.nabranimallaw.org/; accessed October 2, 2010.
Animals and Their Legal Rights. 1985. Washington: Animal Welfare Institute.
APS [American Physiological Society] Guiding Principles for Research Involving Animals and Human Beings. Available at www.the-aps.org/publications/journals/guide.htm; accessed January 24, 2010.
Environmental Policy Tools: A User's Guide. 1995. OTA-ENV-634. Washington: Office of Technology Assessment.
Environmental Regulation: Law, Science and Policy, 4th ed. 2003. Percival R, Schroeder C, Miller A, Leape J. Aspen Publishers. p 128-133.
Society for Neuroscience Policies on the Use of Animals and Humans in Neuroscience Research. Available at http://SFN.org/index.cfm?pagename=guidelinesPolicies_UseOfAnimalsandHumans; accessed January 24, 2010.

US laws and norms related to laboratory animal research. 1999. VandeBerg JL, Williams-Blangero S, Wolfle TL. ILAR J 40:34-37.

Education

AALAS [American Association for Laboratory Animal Science] Technician Training and Certification. Available at www.aalas.org/certification/tech_cert.aspx; accessed May 21, 2010.

CALAS [Canadian Association of Laboratory Animal Science] Registry. Available at www.calas-acsal.org/index.php?option=com_content&task=view&id=24&Itemid=106; accessed August 10, 2010.

Clinical Textbook for Veterinary Technicians, 5th ed. 2002. McCurnin DM, Bassert JM, eds. Philadelphia: WB Saunders.

Education and Training in the Care and Use of Laboratory Animals: A Guide for Developing Institutional Programs. 1991. National Research Council. Washington: National Academy Press.

FELASA recommendations for the accreditation of laboratory animal science education and training. 2002. Nevalainen T, Blom HJ, Guaitani A, Hardy P, Howard BR, Vergara P. Lab Anim 36:373-377.

Guidelines on Institutional Animal User Training. Ottawa. National Institutional Animal User Training Program. 1999. Canadian Council on Animal Care. Available at http://ccac.ca/en/CCAC_Programs/ETCC/Intro-coretopics-Web11.htm; accessed May 15, 2010.

National Need and Priorities for Veterinarians in Biomedical Research. 2004. National Research Council. Washington: National Academies Press.

Perspectives on curriculum needs in laboratory animal medicine. 2009. Turner PV, Colby LA, VandeWoude S, Gaertner DJ, Vasbinder MA. J Vet Med Educ 36:89-99.

The Care and Feeding of an IACUC: The Organization and Management of an Institutional Animal Care and Use Committee. 1999. Podolsky ML, Lucas V, eds. Boca Raton FL: CRC Press.

The IACUC Handbook, 2nd ed. 2006. Silverman J, Sukow MA, Murthy S, eds. Boca Raton FL: CRC Press.

Training and Adult Learning Strategies for the Care and Use of Laboratory Animals. 2007. ILAR J 48(2).

- Formal training programs and resources for laboratory animal veterinarians. 2007. Colby LA, Turner PV, Vasbinder MA. ILAR J 48:143-155.
- Training and adult learning strategies for the care and use of laboratory animals. 2007. Dobrovolny J, Stevens J, Medina LJ. ILAR J 48:75-89.
- Training strategies for animal care technicians and veterinary technical staff. 2007. Pritt S, Duffee N. ILAR J 48:109-119.
- Training strategies for IACUC members and the institutional official. 2007. Greene ME, Pitts ME, James ML. ILAR J 48:132-142.

Monitoring the Care and Use of Animals

A Resource Book for Lay Members of Local Ethical Review Processes, 2nd ed. 2009. Smith JA, Jennings M. West Sussex UK: RSPCA, Research Animals Department. Available at www.rspca.org.uk/ImageLocator/LocateAsset?asset=document&assetId=1232713599355&mode=prd; accessed August 10, 2010.

An IACUC perspective on songbirds and their use as animal models for neurobiological research. 2010. Schmidt MF. ILAR J 51(4):424-430.

Animal Care and Use Committees Bibliography. 1992. Allen T, Clingerman K. Beltsville MD: US Department of Agriculture, National Agricultural Library (Publication #SRB92-16).

Best practices for animal care committees and animal use oversight. 2002. De Haven R. ILAR J 43(Suppl):S59-S62.

Community representatives and nonscientists on the IACUC: What difference should it make? 1999. Dresser R. ILAR J 40:29-33.

Effective animal care and use committees. 1987. In: Orlans FB, Simmonds RC, Dodds WJ, eds. Laboratory Animal Science, Special Issue, January. Published in collaboration with the Scientists Center for Animal Welfare.

Field studies and the IACUC: Protocol review, oversight, and occupational health and safety considerations. 2007. Laber K, Kennedy BW, Young L. Lab Anim 36:27-33.

Guidelines for the veterinary care of laboratory animals. Report of the FELASA/ECLAM/ESLAV Joint Working Group on Veterinary Care. 2008. Lab Anim 42:1-11. Available at http://la.rsmjournals.com/cgi/reprint/42/1/1; accessed May 21, 2010.

Information Resources for Institutional Animal Care and Use Committees 1985-1999. 1999, rev 2000. AWIC Resources Series No. 7. US Department of Agriculture, National Agricultural Library. Available at www.nal.usda.gov/awic/pubs/IACUC/; accessed January 24, 2010.

Institutional Animal Care and Use Committee Guidebook, 2nd ed. 2002. Applied Research Ethics National Association (ARENA), Office of Laboratory Animal Welfare (OLAW), National Institutes of Health. Available at http://grants1.nih.gov/grants/olaw/GuideBook.pdf; accessed January 21, 2010.

Principles and practice in ethical review of animal experiments across Europe: Summary of the report of the FELASA Working Group on Ethical Evaluation of Animal Experiments. 2007. Smith JA, van den Broek FAR, Canto Martorell J, Hackbarth H, Ruksenas O, Zeller W. Lab Anim 41:143-160.

Reference Materials for Members of Animal Care and Use Committees. 1991. Berry DJ. Beltsville MD: US Department of Agriculture, National Agricultural Library (AWIC series #10).

Should IACUCs review scientific merit of animal research projects? 2004. Mann MD, Prentice ED. Lab Anim (NY) 33:26-31.

Supplementary Resources for Lay Members of Local Ethical Review Processes: Projects Involving Genetically Modified Animals. 2004. Lane N, Jennings M. West Sussex UK: RSPCA. Available at www.rspca.org.uk/servlet/BlobServer?blobtable=RSPCABlob&blobcol=urlblob&blobkey=id&blobwhere=1105990223463&blobheader=application/pdf; accessed August 10, 2010.

The IACUC Handbook, 2nd ed. 2007. Silverman J, Suckow MA, Murthy S, eds. Boca Raton FL: CRC Press.

The International Symposium on Regulatory Testing and Animal Welfare: Recommendations on best scientific practices for animal care committees and animal use oversight. 2002. Richmond J, Fletch A, Van Tongerloo R. ILAR J 43(Suppl):S129-S132.

Occupational Health and Safety

Air quality in an animal facility: Particulates, ammonia, and volatile organic compounds. 1996. Kacergis JB, Jones RB, Reeb CK, Turner WA, Ohman JL, Ardman MR, Paigen B. Am Ind Hyg Assoc J 57:634-640.

Allergy to laboratory mice and rats: A review of its prevention, management, and treatment. 1993. Hunskaar S, Fosse RT. Lab Anim 27:206-221.

An overview of the roles and structure of international high-security veterinary laboratories for infectious animal diseases. 1998. Murray PK. Rev Sci Tech Off Int Epiz 17:426-443.

Animal-associated human infections. 1991. Weinberg AN, Weber DJ. Infect Dis Clin North America 5:1-181.

Animal experimentation in level 4 facilities. 2002. Abraham G, Muschilli J, Middleton D. In: Richmond JY, ed. Anthology of Biosafety: BSL-4 Laboratories. Mundelein IL: American Biological Safety Association. p 343-359.

Animal necropsy in maximum containment. 2002. Wilhelmson CL, Jaax NK, Davis K. In: Richmond JY, ed. Anthology of Biosafety: BSL-4 Laboratories. Mundelein IL: American Biological Safety Association. p 361-402.

Association for Assessment and Accreditation of Laboratory Animal Care (AAALAC) International. Position Statement on *Cercopithecine herpesvirus* 1, CHV-1 (Herpesvirus-B). Available at www.aaalac.org/accreditation/positionstatements.cfm; accessed January 24, 2010.

Billions for biodefense: Federal agency biodefense funding, FY2001-FY2005. 2004. Schuler A. Biosecur Bioterror 2:86-96.

Biohazards and Zoonotic Problems of Primate Procurement, Quarantine and Research. 1975. Simmons ML, ed. Cancer Research Safety Monograph Series, vol 2. DHEW Pub. No. (NIH) 76-890. Washington: Department of Health, Education, and Welfare.

Biological Safety Principles and Practices. 2000. Fleming DO, Hunt DL, eds. Washington: ASM Press.

Biosafety in Microbiological and Biomedical Laboratories, 5th ed. 2009. Chosewood CL, Wilson DE, eds. DHHS [Department of Health and Human Services]. Washington: Government Printing Office. Available at http://www.cdc.gov/biosafety/publications/bmbl5/index.htm; accessed July 30, 2010.

Biosafety in the Laboratory: Prudent Practices for Handling and Disposal of Infectious Materials. 1989. National Research Council. Washington: National Academy Press.

Biotechnology Research in an Age of Terrorism. 2004. NRC. Washington: National Academies Press.

Code of Federal Regulations. 1984. Title 40; Part 260, Hazardous Waste Management System: General; Part 261, Identification and Listing of Hazardous Waste; Part 262, Standards Applicable to Generators of Hazardous Waste; Part 263, Standards Applicable to Transporters of Hazardous Waste; Part 264, Standards for Owners and Operators of Hazardous Waste Treatment, Storage, and Disposal Facilities; Part 265, Interim Status Standards for Owners and Operators of Hazardous Waste Treatment, Storage, and Disposal Facilities; and Part 270, EPA-Administered Permit Programs: The Hazardous Waste Permit Program. Washington: Office of the Federal Register. (Part 260 updated April 1994; 261 and 270 updated August 1994; 264 and 265 updated June 1994; 262 and 263 updated 1993)

Evaluation of individually ventilated cage systems for laboratory rodents: Occupational health aspects. 2001. Renstrom A, Bjoring G, Hoglund AU. Lab Anim 35:42-50.

Industrial Biocides. 1988. Payne KR, ed. New York: Wiley.

Infectious Disease Research in the Age of Biodefense. ILAR J 46(1).

Issues related to the use of animals in biocontainment research facilities. Copps J. 2005. ILAR J 46:34-43.

Select agent regulations. Gonder JC. 2005. ILAR J 46:4-7.

Laboratory safety for arboviruses and certain other viruses of vertebrates. 1980. Subcommittee on Arbovirus Safety, American Committee on Arthropod-Borne Viruses. Am J Trop Med Hyg 29:1359-1381.

Mechanism and epidemiology of laboratory animal allergy. 2001. Bush RD. ILAR J 42:4-11.

National Cancer Institute Safety Standards for Research Involving Oncogenic Viruses. 1974. NIH. DHEW Pub. No. (NIH) 78-790. Washington: Department of Health, Education, and Welfare.

NIAID Strategic Plan for Biodefense Research: 2007 Update. National Institute of Allergy and Infectious Diseases, NIH. Available at www3.niaid.nih.gov/topics/BiodefenseRelated/Biodefense/PDF/biosp2007.pdf; accessed January 24, 2010.

NIH Guidelines for Research Involving Recombinant DNA Molecules (NIH Guidelines). 2009. National Institutes of Health. Available at http://oba.od.nih.gov/oba/rac/guidelines_02/NIH_Guidelines_Apr_02.htm; accessed January 24, 2010.

NIH Guidelines for the Laboratory Use of Chemical Carcinogens. 1981. National Institutes of Health. NIH Pub. No. 81-2385. Washington: Department of Health and Human Services.

Occupational Health and Safety in Biomedical Research. 2003. ILAR J 44(1).

An ergonomics process for the care and use of research animals. 2003. Kerst J. ILAR J 44:3-12.

Occupational medicine programs for animal research facilities. 2003. Wald PH, Stave GM. ILAR J 44:57-71.

Occupational Health and Safety in the Care and Use of Nonhuman Primates. 2003. National Research Council. Washington: National Academies Press.

Occupational Health and Safety in the Care and Use of Research Animals. 1997. National Research Council. Washington: National Academy Press.

Personal respiratory protection. 2000. McCullough NV. In: Fleming DO, Hunt DL, eds. Biological Safety Principles and Practices. Washington: ASM Press. p 383-404.

Prudent Practices in the Laboratory: Handling and Disposal of Chemicals. 1995. National Research Council. Washington: National Academy Press.

The Emergence of Zoonotic Diseases. 2002. Burroughs T, Knobler S, Lederberg J, eds. Institute of Medicine. Washington: National Academies Press.

The growing pains of biodefense. 2003. Birmingham K. J Clin Invest 112:970-971.

Zoonoses and Communicable Diseases Common to Man and Animals. 2003. Acha PN, Szyfres B. Washington: Pan American Health Organization.

ENVIRONMENT, HOUSING, AND MANAGEMENT

General References

Biomedical Investigator's Handbook for Researchers Using Animal Models. 1987. Washington: Foundation for Biomedical Research.

Disinfection in Veterinary and Farm Animals Practice. 1987. Linton AH, Hugo WB, Russell AD, eds. Oxford: Blackwell Scientific Publications.

Efficacy of vaporized hydrogen peroxide against exotic animal viruses. 1997. Heckert RA, Best M, Jordan LT, Dulac GC, Eddington DL, Sterritt WG. Appl Environ Micro 63:3916-3918.

Guidelines for the treatment of animals in behavioral research and teaching. 1995. Animal Behavior Society. Anim Behav 49:277-282.

Handbook of Disinfectants and Antiseptics. 1996. Ascenzi JM, ed. New York: Marcel Dekker.

Handbook of Laboratory Animal Science, 2nd ed. 2003. Essential Principles and Practices, vol 1. Boca Raton FL: CRC Press.

IESNA Lighting Handbook, 9th ed. 2000. Illuminating Engineering Society of North America. New York.

Laboratory Animals. 1995. Tuffery AA. London: John Wiley.

Laboratory Animals: An Annotated Bibliography of Informational Resources Covering Medicine, Science (Including Husbandry), Technology. 1971. Cass JS, ed. New York: Hafner Publishing.

Laboratory Animals: An Introduction for New Experimenters. 1987. Tuffery AA, ed. Chichester: Wiley Interscience.
Managing the Laboratory Animal Facility, 2nd ed. 2009. Silverman J, ed. Boca Raton FL: CRC Press.
Microbiological Aspects of Biofilms and Drinking Water. 2000. Percival SL, Walker JT, Hunter PR. Boca Raton FL: CRC Press.
Pheromones and Reproduction in Mammals. 1983. Vandenbergh JG, ed. New York: Academic Press.
Practical Animal Handling. 1991. Anderson RS, Edney ATB, eds. Elmsford NY: Pergamon.
Recent advances in sterilization. 1998. Lagergren ER. J Infect Control (Asia) 1:11-30.
The Experimental Animal in Biomedical Research, vol II: Care, Husbandry, and Well-being: An Overview by Species. 1995. Rollin BE, Kesel ML, eds. Boca Raton FL: CRC Press.
UFAW Handbook on the Care and Management of Laboratory Animals, 7th ed, vol 1: Terrestrial Vertebrates. 1999. Universities Federation for Animal Welfare. Oxford: Blackwell.

Environmental Enrichment

A novel approach for documentation and evaluation of activity patterns in owl monkeys during development of environmental enrichment programs. 2003. Kondo SY, Yudko EB, Magee LK. Contemp Top Lab Anim Sci 42:17-21.
A review of environmental enrichment for pigs housed in intensive housing systems. 2009. van de Weerd HA, Day JEL. Appl Anim Behav Sci 116:1-20.
A review of environmental enrichment strategies for single-caged nonhuman primates. 1989. Fajzi K, Reinhardt V, Smith MD. Lab Anim 18:23-35.
A targeted approach to developing environmental enrichment for two strains of laboratory mice. 2008. Nicol CJ, Brocklebank S, Mendl M, Sherwin C. Appl Anim Behav Sci 110:341-353.
Annotated Bibliography on Refinement and Environmental Enrichment for Primates Kept in Laboratories, 8th ed. 2005. Reinhardt V, Reinhardt A. Washington: Animal Welfare Institute.
Artificial turf foraging boards as environmental enrichment for pair-housed female squirrel monkeys. 2000. Fekete JM, Norcross JL, Newman JD. Contemp Top Lab Anim Sci 39:22-26.
Assessment of the use of two commercially available environmental enrichments by laboratory mice by preference testing. 2005. Van Loo PL, Blom HJ, Meijer MK, Baumans V. Lab Anim 39:58-67.
Behavioural effects of environmental enrichment for individually caged rabbits. 1997. Lidfors L Appl Anim Behav Sci 52:157-169.
Can puzzle feeders be used as cognitive screening instruments? Differential performance of young and aged female monkeys on a puzzle feeder task. 1999. Watson SL, Shively CA, Voytko ML. Am J Primatol 49:195-202.
Effectiveness of video of conspecifics as a reward for socially housed bonnet macaques (*Macaca radiata*). 2004. Brannon E, Andrews M, Rosenblum L. Percept Motor Skills 98(3-1):849-858.
Effects of a cage enrichment program on heart rate, blood pressure, and activity of male Sprague-Dawley and spontaneously hypertensive rats monitored by radiotelemetry. 2005. Sharp J, Azar T, Lawson D. Contemp Top Lab Anim Sci 44:32-40.
Effects of different forms of environmental enrichment on behavioral, endocrinological, and immunological parameters in male mice. 2003. Marashi V, Barnekow A, Ossendorf E, Sachser N. Horm Behav 43:281-292.

Effects of environmental enrichment for mice: Variation in experimental results. 2002. Van de Weerd HA, Aarsen EL, Mulder A, Kruitwagen CL, Hendriksen CF, Baumans V. J Appl Anim Welf Sci 5:87-109.
Effects of environmental enrichment on males of a docile inbred strain of mice. 2004. Marashi V, Barnekow A, Sachser N. Physiol-Behav 82:765-776.
Effects of puzzle feeders on pathological behavior in individually housed rhesus monkeys. 1998. Novak MA, Kinsey JH, Jorgensen MJ, Hazen TJ. Am J Primatol 46:213-227.
Enriching the environment of the laboratory cat. 1995. AWIC Resource Series, No. 2: Environmental Enrichment Information Resources for Laboratory Animals—Birds, Cats, Dogs, Farm Animals, Ferrets, Rabbits, and Rodents. McCune S. Beltsville MD: AWIC. p 27-33.
Enrichment and aggression in primates. 2006. Honess PE, Marin CM. Neurosci Biobehav Rev 30:413-436.
Enrichment of laboratory caging for rats: A review. 2004. Patterson-Kane EF. Anim Welf 13:209-214.
Enrichment strategies for laboratory animals. 2005. ILAR J 46(2).
Enrichment and nonhuman primates: "First, do no harm." 2005. Nelson RJ, Mandrell TD. ILAR J 46:171-177.
Environmental enrichment for laboratory rodents and rabbits: Requirements of rodents, rabbits, and research. 2005. Baumans V. ILAR J 46:162-170.
Environmental enrichment for nonhuman primates. Lutz CK, Novak MA. 2005. ILAR J 46:178-191.
Environmental Enrichment Information Resources for Nonhuman Primates: 1987-1992. 1992. National Agricultural Library, National Library of Medicine, and Primate Information Center. Beltsville MD: National Agricultural Library.
Environmental Enrichment for Caged Rhesus Macaques, 2nd ed. 2001. Reinhardt V, Reinhardt A. Washington: Animal Welfare Institute.
Environmental Enrichment for Captive Animals. 2003. Young RJ. Oxford: Blackwell Science.
Environmental enrichment for nonhuman primates: An experimental approach. 2000. de Rosa C, Vitale A, Puopolo M. In: Progress in the Reduction, Refinement and Replacement of Animal Experimentation: Proceedings of the 3rd World Congress on Alternatives and Animal Use in the Life Sciences, Bologna, Italy. Elsevier. p 1295-1304.
Environmental enrichment in mice decreases anxiety, attenuates stress responses and enhances natural killer cell activity. 2004. Benaroya-Milshtein N, Hollander N, Apter A, Kukulansky T, Raz N, Wilf A, Yaniv I, Pick CG. Eur J Neurosci 20:1341-1347.
Environmental enrichment lowers stress-responsive hormones in singly housed male and female rats. 2003. Belz EE, Kennell JS, Czambel RK, Rubin RT, Rhodes ME. Pharmacol Biochem Behav 76:481-486.
Environmental enrichment may alter the number of rats needed to achieve statistical significance. 1999. Eskola S, Lauhikari M, Voipio HM, Laitinen M, Nevalainen T. Scand J Lab Anim Sci 26:134-144.
Environmental enrichment of laboratory animals used in regulatory toxicology studies. 1999. Dean SW. Lab Anim 33:309-327.
Environmental enrichment of nonhuman primates, dogs and rabbits used in toxicology studies. 2003. Bayne KA. Toxicol Pathol 31(Suppl):132-137.
Environmental enrichment options for laboratory rats and mice. 2004. Key D. Lab Anal 33:39-44.
Evaluation of objects and food for environmental enrichment of NZW rabbits. 2001. Harris LD. Contemp Top Lab Anim Sci 40:27-30.
Guidelines for developing and managing an environmental enrichment program for nonhuman primates. 1991. Bloomsmith MA, Brent LY, Schapiro SJ. Lab Anim Sci 41:372-377.

Housing, Care and Psychological Well-Being of Captive and Laboratory Primates. 1989. Segal EF, ed. Park Ridge NJ: Noyes Publications.
Incorporation of an enrichment program into a study protocol involving long-term restraint in macaques. 2002. McGuffey LH, McCully CL, Bernacky BJ, Blaney SM. Lab Anim 31:37-39.
Monkey behavior and laboratory issues. 1998. Bayne KA, Novak M, eds. Lab Anim Sci 41:306-359.
Nonhuman Primate Management Plan. 1991. Office of Animal Care and Use, NIH. Available at http://oacu.od.nih.gov/regs/primate/primex.htm; accessed January 24, 2010.
Psychological Well-Being of Nonhuman Primates. 1998. National Research Council. Washington: National Academy Press.
Psychological well-being of nonhuman primates: A brief history. 1999. Wolfle TL. J Appl Anim Welf Sci 2:297-302.
Rodent enrichment dilemmas: The answers are out there! 2004. Hawkins P, Jennings M. Anim Technol Welf 3:143-147.
Shelter enrichment for rats. 2003. Patterson-Kane EG. Contemp Top Lab Anim Sci 42:46-48.
Short-term effects of an environmental enrichment program for adult cynomolgus monkeys. 2002. Turner PV, Grantham LE II. Contemp Top Lab Anim Sci 41:13-17.
Social enhancement for adult nonhuman primates in research laboratories: A review. 2000. Reinhardt V, Reinhardt A. Lab Anim 29:34-41.
The effect of non-nutritive environmental enrichment on the social behavior of group-housed cynomolgus macaques (*Macaca fascicularis*). 2003. Kaplan J, Ayers M, Phillips M, Mitchell C, Wilmoth C, Cairnes D, Adams M. Contemp Top Lab Anim Sci 42:117.
The need for responsive environments. 1990. Markowitz H, Line S. In: Rollin BE, Kesel ML, eds. The Experimental Animal in Biomedical Research, vol I: A Survey of Scientific and Ethical Issues for Investigators. Boca Raton FL: CRC Press. p 153-172.
Through the Looking Glass: Issues of Psychological Well-Being in Captive Nonhuman Primates. 1991. Novak M, Petto AJ, eds. Washington: American Psychological Association.

Genetics and Genetically Modified Animals

A Primer of Population Genetics, 3rd ed. 2000. Hartl DL. Sunderland MA: Sinauer Associates.
Animal production and breeding methods. Festing MFW, Peters AG. 1999. In: Poole T, ed. The UFAW Handbook on the Care and Management of Laboratory Animals, 7th ed, vol 1. Oxford: Blackwell Sciences. p 28-44.
Assessing the welfare of genetically altered mice. 2006. Wells DJ. Lab Anim 39:314-320.
Behavioral phenotyping of transgenic and knockout mice: Experimental design and evaluation of general health, sensory functions, motor abilities, and specific behavioral tests. 1999. Crawley JN. Brain Res 835:18-26.
Effective population size, genetic variation, and their use in population management. 1987. Lande R, Barrowclough G. In: Soule M, ed. Viable Populations for Conservation. Cambridge, UK: Cambridge University Press. p 87-123.
Genetically Engineered Mice Handbook. 2006. Sundberg JP, Ichiki T, eds. Boca Raton FL: CRC Press.
Genetically modified animals: What welfare problems do they face? 2003. Buehr M, Hjorth JP. J Appl Anim Welf Sci 6:319-338.
Genetics and Probability in Animal Breeding Experiments. 1981. Green EL. New York: Oxford University Press.
Gnotobiology and breeding techniques. 2004. Hardy P. In: Hedrich H, ed. The Laboratory Mouse. London: Elsevier. p 409-433.

Joint Working Group on Refinement: Refinement and reduction in production of genetically modified mice. 2003. Sixth report of the BVAAWF/FRAME/RSPCA/UFAW Joint Working Group on Refinement. Lab Anim 37(Suppl 1):S1-S49. Available at http://la.rsmjournals.com/cgi/reprint/37/suppl_1/1.pdf; accessed August 10, 2010.
Laboratory animal genetics and genetic quality control. 2003. Festing MF. In: Hau J, Van Hoosier GL Jr, eds. Handbook of Laboratory Animal Science, 2nd ed, vol 1. Boca Raton FL: CRC Press. p 173-203.
Making better transgenic models: Conditional, temporal, and spatial approaches. 2005. Ristevski S. Mol Biotechnol 29:153-163.
Mouse functional genomics requires standardization of mouse handling and housing conditions. 2004. Champy MF, Selloum M, Piard L, Zeitler V, Caradec C, Chambon P, Auwerx J. Mamm Genome 15:768-783.
Persistent transmission of mouse hepatitis virus by transgenic mice. 2001. Rehg JE, Blackman MA, Toth LA. Comp Med 51:369-374.
Projects Involving Genetically Modified Animals. 2004. Lane N, Jennings M. RSPCA, Research Animals Department. Available at www.rspca.org.uk/ImageLocator/LocateAsset?asset=documentandassetId=1232712279548andmode=prd; accessed January 24, 2010.
Reproduction and Breeding Techniques for Laboratory Animals. 1970. Hafez ESE, ed. Philadelphia: Lea and Febiger.
Research-oriented genetic management of nonhuman primate colonies. 1993. Williams-Blangero S. Lab Anim Sci 43:535-540.
Rules and Guidelines for Nomenclature of Mouse and Rat Strains, rev. 2007. International Committee on Standardized Genetic Nomenclature for Mice. Available at www.informatics.jax.org/mgihome/nomen/2007_strains.shtml; accessed May 15, 2010.
Standardized nomenclature for transgenic animals. 1992. ILAR News 34:45-52.
The use of genetic "knockout" mice in behavioral endocrinology research. 1997. Nelson RJ. Horm Behav 31:188-196.
Transgenic animal technology: Alternatives in genotyping and phenotyping. 2003. Pinkert CA. Comp Med 53:126-139.
Transgenic gene knock-outs: Functional genomics and therapeutic target selection. 2000. Harris S, Ford SM. Pharmacogenomics 1:433-443.

Species-Specific References—Environment, Housing, and Management

Agricultural Animals

Behavior of Domestic Animals. 1985. Hart BL. New York: WH Freeman.
Cattle: Good Practice for Housing and Care, 1st ed. 2008. RSPCA, Research Animals Department. Available at www.rspca.org.uk/servlet/BlobServer?blobtable=RSPCABlob&blobcol=urlblob&blobkey=id&blobwhere=1220375292149&blobheader=application/pdf; accessed August 6, 2010.
Comfortable quarters for cattle in research institutions. 2002. Reinhardt V, Reinhardt A. In: Comfortable Quarters for Laboratory Animals, 9th ed. Washington: Animal Welfare Institute. p 89-95.
Comfortable quarters for horses in research institutions. 2002. Houpt KA, Ogilvie-Graham TS. In: Comfortable Quarters for Laboratory Animals, 9th ed. Washington: Animal Welfare Institute. p 96-100.
Comfortable quarters for pigs in research institutions. 2002. Grandin T. In: Comfortable Quarters for Laboratory Animals, 9th ed. Washington: Animal Welfare Institute. p 78-82.

Comfortable quarters for sheep in research institutions. 2002. Reinhardt V. In: Comfortable Quarters for Laboratory Animals, 9th ed. Washington: Animal Welfare Institute. p 83-88.

Dukes' Physiology of Domestic Animals, 12th ed. 2004. Reece WO, ed. Ithaca NY: Cornell University Press.

Essentials of Pig Anatomy. 1982. Sack WO. Ithaca NY: Veterinary Textbooks.

Farm Animal Housing and Welfare. 1983. Baxter SH, Baxter MR, MacCormack JAC, eds. Boston: Nijhoff.

Farm Animals and the Environment. 1992. Phillips C, Piggins D, eds. Wallingford UK: CAB International.

Indicators Relevant to Farm Animal Welfare. 1983. Smidt D, ed. Boston: Nijhoff.

Management and Welfare of Farm Animals, 4th ed. 1999. Ewbank R, Kim-Madslien F, Hart CB. Universities Federation for Animal Welfare (out of print; 5th edition in preparation).

Miniature pet pigs. 1999. Van Metre DC, Angelos SM. Vet Clin North America Exot Anim Pract 2:519-537.

Pigs: Good Practice for Housing and Care, 2nd ed. 2008. RSPCA, Research Animals Department. Available at http://content.www.rspca.org.uk/cmsprd/Satellite?blobcol=urldata&blobheader=application%2Fpdf&blobkey=id&blobnocache=false&blobtable=MungoBlobs&blobwhere=1232988745105&ssbinary=true; accessed January 24, 2010.

Policy on the care and use of sheep for scientific purposes based on good practice. 2005. Mellor DJ, Hemsworth PH. Victoria, Australia: Monash University and the Animal Welfare Science Centre.

Reproduction in Farm Animals. 1993. Hafez ESE. Philadelphia: Lea and Febiger.

Responsiveness, behavioural arousal and awareness in fetal and newborn lambs: Experimental, practical and therapeutic implications. 2003. Mellor DJ, Gregory NG. NZ Vet J 51:2-13.

Restraint of Domestic Animals. 1991. Sonsthagen TF. Goleta CA: American Veterinary Publications.

Ruminants: Cattle, Sheep, and Goats. 1974. Guidelines for the Breeding, Care and Management of Laboratory Animals. National Research Council. Washington: National Academy of Sciences.

Sheep: Good Practice for Housing and Care, 2nd ed. 2008. RSPCA, Research Animals Department. Available at http://content.www.rspca.org.uk/cmsprd/Satellite?blobcol=urldata&blobheader=application%2Fpdf&blobkey=id&blobnocache=false&blobtable=MungoBlobs&blobwhere=1232988745203&ssbinary=true; accessed January 24, 2010.

Swine as Models in Biomedical Research. 1992. Swindle MM. Ames: Iowa State University Press.

Swine in Cardiovascular Research. 1986. Stanton HC, Mersmann HJ. Boca Raton FL: CRC Press.

Taming and training of pregnant sheep and goats and of newborn lambs, kids and calves before experimentation. 2004. Mellor DJ. ALTA 32(S1):143-146.

The Biology of the Pig. 1978. Pond WG, Houpt KA. Ithaca NY: Comstock Publishing.

The Calf: Management and Feeding, 5th ed. 1990. Roy JHB. Boston: Butterworths.

The Sheep as an Experimental Animal. 1983. Heckler JF. New York: Academic Press.

Welfare concerns for farm animals used in agriculture and biomedical research and teaching. 1994. Swanson JC. Animal Welfare Information Center Newsletter, Special Issue: Farm Animals in Research and Teaching 5 (01), USDA.

Amphibians, Reptiles, and Fish

Amphibian Medicine and Captive Husbandry. 2001. Wright K, Whitaker BR. Malabar FL: Krieger Publishing Company.

Amphibians. 1999. Halliday TR. In: Poole TB, ed. UFAW Handbook on the Care and Management of Laboratory Animals, 7th ed, vol 2. Oxford: Blackwell.

Amphibious and aquatic vertebrates and advanced invertebrates. 1999. In UFAW Handbook on the Care and Management of Laboratory Animals, 7th ed, vol 2: Poole TB, ed. Oxford: Blackwell.

Anthropomorphism and "mental welfare" of fishes. 2007. Rose JD. Dis Aquat Org 75:139-154.

Artificial Seawaters: Formulas and Methods. 1985. Bidwell JP, Spotte S. Boston: Jones and Bartlett.

Captive Seawater Fishes: Science and Technology. 1992. Spotte SH. Wiley-Interscience.

CCAC Guidelines on the Care and Use of Fish in Research, Teaching, and Testing. 2005. Canadian Council on Animal Care. Ottawa. Available at www.ccac.ca/en/CCAC_Programs/Guidelines_Policies/GDLINES/Fish/Fish_Guidelines_English.pdf; accessed January 24, 2010.

Comfortable quarters for amphibians and reptiles in research institutions. 2002. Kreger MD. In: Comfortable Quarters for Laboratory Animals, 9th ed. Washington: Animal Welfare Institute. p 109-114.

Current issues in fish welfare. 2006. Huntingford FA, Adams C, Braithwaite VA, Kadri S, Pottinger TG, Sandøe P, Turnbull JF. J Fish Biol 68:332-372.

Enrichment for a captive environment: The *Xenopus laevis*. 2004. Brown MJ, Nixon RM. Anim Technol Welf 3:87-95.

Fish, Amphibians, and Reptiles. 1995. ILAR J 37(4).

Guidelines for the care and use of fish in research. 1995. Detolla LJ, Srinivas S, Whitaker BR, Andrews C, Hecker B, Kane AS, Reimschuessel R. ILAR J 37:159-173.

Fish Models in Biomedical Research. 2001. ILAR J 42(4).

A fish model of renal regeneration and development. 2001. Reimschuessel R. ILAR J 42:285-291, 305-308.

Development of sensory systems in zebrafish (*Danio rerio*). 2001. Moorman SJ. ILAR J 42:292-298.

Mechanistic considerations in small fish carcinogenicity testing. 2001. Law JM. ILAR J 42:274-284.

Transgenic fish as models in environmental toxicology. 2001. Winn RN. ILAR J 42:322-329.

Fish Pathology, 2nd ed. 1989. Roberts RJ, ed. London: Saunders.

Frogs and toads as experimental animals. 1999. Tyler MJ. ANZCCART News 12(1) Insert.

Guidance on the Housing and Care of the African Clawed Frog. 2005. Reed BT. RSPCA, Research Animals Department. Available at www.rspca.org.uk/ImageLocator/LocateAsset?asset=document&assetId=1232712646624&mode=prd; accessed August 10, 2010.

Guidelines for the health and welfare monitoring of fish used in research. 2006. Johansen R, Needham JR, Colquhoun DJ, Poppe TT, Smith AJ. Lab Anim 40:323-240.

Guidelines for the Use of Fishes in Research. 2004. Use of Fishes in Research Committee. Joint publication of the American Fisheries Society, the American Society of Ichthyologists and Herpetologists, and the American Institute of Fisheries Research Biologists. Available at www.fisheries.org/afs/docs/policy_16.pdf; accessed January 24, 2010.

Guidelines for the Use of Live Amphibians and Reptiles in Field and Laboratory Research, 2nd ed, rev. 2004. Herpetological Animal Care and Use Committee, American Society of Ichthyologists and Herpetologists. Available at www.asih.org/files/hacc-final.pdf; accessed January 24, 2010.

Housing and husbandry of *Xenopus* for oocyte production. 2003. Schultz TW, Dawson DA. Lab Anim 32:34-39.

Important ethological and other considerations of the study and maintenance of reptiles in captivity. 1990. Warwick C. Appl Anim Behav Sci 27:363-366.

Information Resources for Reptiles, Amphibians, Fish, and Cephalopods Used in Biomedical Research. Berry DJ, Kreger MD, Lyons-Carter JL. 1992. Beltsville MD: USDA National Library Animal Welfare Information Center.

Key issues concerning environmental enrichment for laboratory-held fish species. 2009. Williams TD. Lab Anim 43:107-120.

Laboratory Anatomy of the Turtle. 1955. Ashley LM. Dubuque IA: WC Brown.

Nontraditional Animal Models for Biomedical Research. 2004. ILAR J 45(1).

The green anole (*Anolis carolinensis*): A reptilian model for laboratory studies of reproductive morphology and behavior. 2004. Lovern MB, Holmes MM, Wade J. ILAR J 45:54-64.

The male red-sided garter snake (*Thamnophis sirtalis parietalis*): Reproductive pattern and behavior. 2004. Krohmer RW. ILAR J 45:65-74.

Tracing the evolution of brain and behavior using two related species of whiptail lizards: *Cnemidophorus uniparens* and *Cnemidophorus inornatus*. 2004. Woolley SC, Sakata JT, Crews D. ILAR J 45:46-53.

Safeguarding the many guises of farmed fish welfare. 2007. Turnbull JF, Kadri S. Dis Aquat Org 75:173-182.

Stress and the welfare of cultured fish. 2004. Conte FS. Appl Anim Behav Sci 86:205-223.

The Care and Use of Amphibians, Reptiles, and Fish in Research. 1992. Schaeffer DO, Kleinow KM, Krulisch L, eds. Proceedings from a SCAW/LSU-SVM-sponsored conference, April 8-9, 1991, New Orleans. Greenbelt MD: Scientists Center for Animal Welfare.

The Handbook of Experimental Animals: The Laboratory Fish. 2000. Ostrander GK, Bullock GR, Bunton T, eds. Academic Press.

The Laboratory Xenopus. 2009. Green SL. Boca Raton FL: CRC Press.

The Zebrafish Book, 5th ed: A Guide for the Laboratory Use of Zebrafish (*Danio rerio*). 2007. Westerfield M. Eugene: University of Oregon Press.

Use of Amphibians in the Research, Laboratory, or Classroom Setting. 2007. ILAR J 48(3).

Birds

Avian and Exotic Animal Hematology and Cytology. 2007. Campbell TW, Ellis CK. Oxford: Wiley-Blackwell.

Birds as Animal Models in the Behavioral and Neural Sciences. 2010. ILAR J 51(4).

A social ethological perspective applied to care of and research on songbirds. 2010. White DJ. ILAR J 51(4):387-393.

Guidelines and ethical considerations for housing and management of psittacine birds used for research. Kalmar ID, Janssens GPJ, Moons CPH. ILAR J 51(4):409-423.

The use of passerine bird species in laboratory research: Implications of basic biology for husbandry and welfare. 2010. Bateson M, Feenders G. ILAR J 51(4):394-408.

Chicken welfare is influenced more by housing conditions than by stocking density. 2004. Dawkins MS, Donnelly CA, Jones TA. Nature 427:342-344.

Comfortable quarters for chickens in research institutions. 2002. Fölsch DW, Höfner M, Staack M, Trei G. In: Comfortable Quarters for Laboratory Animals, 9th ed. Washington: Animal Welfare Institute. p 101-108.

Domestic Fowl: Good Practice for Housing and Care, 2nd ed. 2008. RSPCA, Research Animals Department. Available at http://content.www.rspca.org.uk/cmsprd/Satellite?blobcol=urldata&blobheader=application%2Fpdf&blobkey=id&blobnocache=false&blobtable=MungoBlobs&blobwhere=1232988515260&ssbinary=true; accessed January 24, 2010.
Ducks and Geese: Good Practice for Housing and Care, 2nd ed. 2008. RSPCA, Research Animals Department. Available at http://content.www.rspca.org.uk/cmsprd/Satellite?blobcol=urldata&blobheader=application%2Fpdf&blobkey=id&blobnocache=false&blobtable=MungoBlobs&blobwhere=1232988515191&ssbinary=true; accessed January 24, 2010.
Guidelines to the Use of Wild Birds in Research. 2010. Fair J, Paul E, Jones J, eds. Washington: Ornithological Council. Available at http://www.nmnh.si.edu/BIRDNET/guide/index.html; accessed August 20, 2010.
Laboratory Animal Management: Wild Birds. 1977. National Research Council. Washington: National Academy of Sciences.
Laboratory Birds: Refinements in Husbandry and Procedures. 2003. Fifth report of the BVAAWF/FRAME/RSPCA/UFAW Joint Working Group on Refinement. Lab Anim 35(Suppl 1):S1-S163. Available at www.rspca.org.uk/ImageLocator/LocateAsset?asset=document&assetId=1232712322818&mode=prd; accessed August 10, 2010.
Manual of Ornithology: Avian Structure and Function. 2007. Proctor NS, Lynch PJ. In: Thomas NJ, Hunter DB, Atkinson CT, eds. Infectious Diseases of Wild Birds. Oxford: Wiley-Blackwell.
Physiology and Behavior of the Pigeon. 1983. Abs M, ed. London: Academic Press.
Pigeons: Good Practice for Housing and Care, 2nd ed. 2008. RSPCA, Research Animals Department. Available at http://content.www.rspca.org.uk/cmsprd/Satellite?blobcol=urldata&blobheader=application%2Fpdf&blobkey=id&blobnocache=false&blobtable=MungoBlobs&blobwhere=1232988515283&ssbinary=true; accessed January 24, 2010.
Quail: Good Practice for Housing and Care, 3rd ed. 2008. RSPCA, Research Animals Department. Available at http://content.www.rspca.org.uk/cmsprd/Satellite?blobcol=urldata&blobheader=application%2Fpdf&blobkey=id&blobnocache=false&blobtable=MungoBlobs&blobwhere=1232988515308&ssbinary=true; accessed January 24, 2010.
Sturkie's Avian Physiology, 5th ed. 1999. Whittow GC, ed. San Diego and London: Academic Press.
The domestic chicken. 1996. Glatz P. ANZCCART News 9(2) Insert.
The Pigeon. 1974 (reprinted 1981). Levi WM. Sumter SC: Levi Publishing.
Use and husbandry of captive European starlings (*Sturnus vulgaris*) in scientific research: A review and current best practice. 2008. Asher L, Bateson M. Lab Anim 42:127-139.
Zebra Finches: Good Practice for Housing and Care, 2nd ed. 2008. RSPCA, Research Animals Department. Available at http://content.www.rspca.org.uk/cmsprd/Satellite?blobcol=urldata&blobheader=application%2Fpdf&blobkey=id&blobnocache=false&blobtable=MungoBlobs&blobwhere=1232988515331&ssbinary=true; accessed January 24, 2010.

Cats and Dogs

A comparison of tethering and pen confinement of dogs. 2001. Yeon SC, Golden G, Sung W, Erb HN, Reynolds AJ, Houpt KA. J Appl Anim Welf Sci 4:257-270.
A practitioner's guide to working dog welfare. 2009. Rooney N, Gaines S, Hiby E. J Vet Behav 4:127-134.
Behavioural and physiological correlates of stress in laboratory cats. 1993. Carlstead K, Brown JL, Strawn W. Appl Anim Behav Sci 38:143-158.
Canine Anatomy: A Systematic Study. 1986. Adams DR. Ames: Iowa State University Press.

Comfortable quarters for cats in research institutions. 2002. Rochlitz I. In: Comfortable Quarters for Laboratory Animals, 9th ed. Washington: Animal Welfare Institute. p 50-55.

Comfortable quarters for dogs in research institutions. 2002. Hubrecht R. In: Comfortable Quarters for Laboratory Animals, 9th ed. Washington: Animal Welfare Institute. p 56-64.

Dogs: Good Practice for Housing and Care, 2nd ed. 2008. RSPCA, Research Animals Department. Available at http://content.www.rspca.org.uk/cmsprd/Satellite?blobcol=urldata&blobheader=application%2Fpdf&blobkey=id&blobnocache=false&blobtable=MungoBlobs&blobwhere=1232988745038&ssbinary=true; accessed January 24, 2010.

Feline behavioral guidelines from the American Association of Feline Practitioners. 2005. Overall KL, Rodan I, Beaver BV, Careny H, Crowell-Davis S, Hird N, Kudrak S, Wexler-Mitchell E, Zicker S. JAVMA 227:70-84.

Joint Working Group on Refinement: Husbandry refinements for rats, mice, dogs and non-human primates used in telemetry procedures. Seventh report of the BVAAWF/FRAME/RSPCA/UFAW Joint Working Group on Refinement, Part B. 2004. Hawkins P, Morton DB, Beyan R, Heath K, Kirkwood J, Pearce P, Scott L, Whelan G, Webb A. Lab Anim 38:1-10. Available at www.rspca.org.uk/ImageLocator/LocateAsset?asset=document&assetId=1232712323251&mode=prd; accessed August 10, 2010.

Joint Working Group on Refinement: Refining dog husbandry and care. 2004. Eighth report of the BVAAWF/FRAME/RSPCA/UFAW Joint Working Group on Refinement. Lab Anim 38 (Suppl 1):S1-S94. Available at www.rspca.org.uk/ImageLocator/LocateAsset?asset=document&assetId=1232712322899&mode=prd; accessed August 10, 2010.

Laboratory Animal Management: Cats. 1978. ILAR News 21(3):C1-C20.

Laboratory Animal Management: Dogs. 1994. National Research Council. Washington: National Academy Press.

Method for long-term cerebrospinal fluid collection in the conscious dog. 1998. Wilsson Rahmberg M, Olovson SG, Forshult E. J Invest Surg 11:207-214.

Miller's Anatomy of the Dog, 3rd ed. 1993. Evans HE. Philadelphia: WB Saunders.

Modern concepts of socialisation for dogs: Implications for their behaviour, welfare and use in scientific procedures. 2004. Boxall J, Heath S, Brautigam J. ALTA 32(Suppl 2):81-93.

Recommendations for the housing and care of domestic cats in laboratories. 2000. Rochlitz I. Lab Anim 34:1-9.

Results of the Survey of Dog Accommodation and Care. 1998. The Animal Procedures Committee. In the Report of the Animal Procedures Committee. London: The Stationary Office. Available at http://apc.homeoffice.gov.uk/reference/ar98.pdf; accessed January 24, 2010.

The Beagle as an Experimental Dog. 1970. Andersen AC, ed. Ames: Iowa State University Press.

The Canine as a Biomedical Research Model: Immunological, Hematological, and Oncological Aspects. 1980. Shifrine M, Wilson FD, eds. UC Davis, Laboratory for Energy-related Research. (May be ordered as DOE/TIC-10191 from National Technical Information Service, www.ntis.gov/search/product.aspx?ABBR=DOETIC10191.)

The domestic cat. 1999. McCune S. In: Poole T, ed. UFAW Handbook on the Care and Management of Laboratory Animals. Oxford: Blackwell. p 445-463.

The effect of housing and handling practices on the welfare, behaviour and selection of domestic cats (*Felis silvestris catus*) by adopters in an animal shelter. 2006. Gourkow N, Fraser D. Anim Welf 15:371-377.

The laboratory cat. 1995. James AE. ANZCCART News 8(1) Insert.

Exotic, Wild, and Zoo Animals

Animal Training: Successful Animal Management Through Positive Reinforcement. 1999. Ramirez K. Chicago: Shedd Aquarium Society.

Biology, Medicine, and Surgery of South American Wild Animals. 2001. Fowler ME, Cubas ZS. Ames: Iowa State University Press

CCAC Guidelines on the Care and Use of Wildlife. 2003. Canadian Council on Animal Care. Available at www.ccac.ca/en/CCAC_Programs/Guidelines_Policies/GDLINES/Wildlife/Wildlife.pdf; accessed January 24, 2010.

Environmental Enrichment for Captive Animals (UFAW Animal Welfare). 2003. Young RJ. Oxford: Wiley-Blackwell.

Essentials of Disease in Wild Animals. 2005. Wobeser GA. Oxford: Wiley-Blackwell.

Exotic and laboratory animals. 2008. In: Kahn CM, ed. Merck Veterinary Manual. Whitehouse Station NJ: Merck and Co.

Exotic Animal Formulary, 3rd ed. 2004. Carpenter JW. Philadelphia: WB Saunders.

Field studies and the IACUC: Protocol review, oversight, and occupational health and safety considerations. 2007. Laber K, Kennedy BW, Young L. Lab Anim 36:27-33.

Guidelines for the Capture, Handling, and Care of Mammals. 1998. The American Society of Mammalogists. Available at www.mammalsociety.org/committees/commanimalcareuse/98acucguidelines.pdf; accessed January 24, 2010.

Guidelines of the American Society of Mammologists for the Use of Wild Mammals in Research. 2007. Gannon WL, Sikes RS, and Animal Care and Use Committee of the American Society of Mammalogists. J Mammal 88:809-823. Available at www.mammalogy.org/committees/commanimalcareuse/ASM%20Animal%20Guidelines.pdf; accessed January 24, 2010.

Information Resources on Big Cats. 2008. Crawford RL. Beltsville MD: USDA National Agricultural Library Animal Welfare Information Center.

Restraint and Handling of Wild and Domestic Animals. 2008. Fowler E. Oxford: Wiley-Blackwell.

Techniques for Wildlife Investigations and Management, 6th ed. 2005. Braun CE, ed. Bethesda MD: The Wildlife Society.

Second Nature: Environmental Enrichment for Captive Animals. 2007. Shepherdson DJ, Mellen JD, Hutchins M. Washington: Smithsonian Institution Press.

Wild Mammals in Captivity: Principles and Techniques. 1997. Kleiman DG, Allen ME, Thompson KV, Lumpkin S. Chicago: University of Chicago Press.

Wildlife as source of zoonotic infections. 2004. Kruse H, Kirkemo A-M, Handeland K. Emerg Infect Dis 10:2067-2072.

Nonhuman Primates

A study of behavioural responses of non-human primates to air transport and re-housing. 2004. Honess PE, Johnson PJ, Wolfensohn SE. Lab Anim 38:119-132.

Aging in Nonhuman Primates. 1979. Bowden DM, ed. New York: Van Nostrand.

Behavior and Pathology of Aging in Rhesus Monkeys. 1985. Davis RT, Leathrus CW, eds. New York: Alan R. Liss.

Best practice in the accommodation and care of primates used in scientific procedures. 2004. MRC Ethics Guide. London: Medical Research Council.

Cage sizes for tamarins in the laboratory. 2004. Prescott MJ, Buchanan-Smith HM. Anim Welf 13:151-157.

Captivity and Behavior: Primates in Breeding Colonies, Laboratories and Zoos. 1979. Erwin J, Maple TL, Mitchell G, eds. New York: Van Nostrand.

Care and Management of Chimpanzees (*Pan troglodytes*) in Captive Environments. 1992. Fulk R, Garland C, eds. Asheboro: North Carolina Zoological Society.

Comfortable quarters for nonhuman primates in research institutions. 2002. Reinhardt V. In: Comfortable Quarters for Laboratory Animals, 9th ed. Washington: Animal Welfare Institute. p 65-77.
Creating housing to meet the behavioral needs of long-tailed macaques. 2008. Waitt CD, Honess PE, Bushmitz M. Lab Primate Newsl 47(4):1-5. Available at www.brown.edu/Research/Primate/lpn47-4.html; accessed January 24, 2010.
Handbook of Primate Husbandry and Welfare. 2005. Wolfensohn S, Honess P. Ames IA: Blackwell Publishing.
Handbook of Squirrel Monkey Research. 1985. Rosenblum LA, Coe CL, eds. New York: Plenum Press.
Housing and care of monkeys and apes in laboratories: Adaptations allowing essential species-specific behaviour. 2002. Roder EL, Timmermans PJ. Lab Anim 36:221-242.
Implementation of permanent group housing for cynomolgus macaques on a large scale for regulatory toxicology studies. 2008. Kelly J. ALTEX 14:107-110.
IPS International Guidelines for the Acquisition, Care, and Breeding of Nonhuman Primates, 2nd ed. 2007. Captive Care Committee, International Primatological Society. Available at www.internationalprimatologicalsociety.org/docs/IPS_International_Guidelines_for_the_Acquisition_Care_and_Breeding_of_Nonhuman_Primates_Second_Edition_2007.pdf; accessed January 24, 2010.
Joint Working Group on Refinement: Refinements in husbandry, care and common procedures for non-human primates. Ninth report of the BVAAWF/FRAME/RSPCA/UFAW Joint Working Group on Refinement. 2009. Jennings M, Prescott MJ. Lab Anim 43:S1-S47.
Joint Working Group on Refinement: Husbandry refinements for rats, mice, dogs and non-human primates used in telemetry procedures. Seventh report of the BVAAWF/FRAME/RSPCA/UFAW Joint Working Group on Refinement, Part B. 2004. Hawkins P, Morton DB, Beyan R, Heath K, Kirkwood J, Pearce P, Scott L, Whelan G, Webb A. Lab Anim 38:1-10. Available at www.rspca.org.uk/ImageLocator/LocateAsset?asset=document&assetId=1232712323251&mode=prd; accessed August 10, 2010.
Laboratory Animal Management: Nonhuman Primates. 1980. ILAR News 23(2-3):P1-P44.
Living New World Monkeys (*Platyrrhini*). 1977. Hershkovitz P. Chicago: University of Chicago Press.
NC3Rs Guidelines: Primate Accommodation, Care and Use. 2006. London: National Centre for the Replacement, Refinement and Reduction of Animals in Research. Available at www.nc3rs.org.uk/downloaddoc.asp?id=418&page=277&skin=0; accessed January 24, 2010.
Nonhuman Primates in Biomedical Research: Biology and Management. Bennett BT, Abee CR, Henrickson R, eds. 1995. New York: Academic Press.
Postsurgical pairing: A discussion by the Refinement and Enrichment Forum. 2006. Van Loo P, Skoumbourdis E, Reinhardt V. Anim Technol Welf 5:17-19.
Primate housing: A new approach. 2003. Rudling W. Anim Technol Welf 2:143-150.
Primate sensory capabilities and communication signals: Implications for care and use in the laboratory. 2006. Prescott M. NC3Rs #4: Primate Senses and Communication. Available at http://nc3rs.tnllive.co.uk/news.asp?id=187; accessed August 10, 2010.
Refinement of the use of non-human primates in scientific research, part II: Housing, husbandry and acquisition. 2006. Rennie AE, Buchanan-Smith HM. Anim Welf 15:215-238.
Safe Pair Housing of Macaques. 2008. Carlson J. Washington: Animal Welfare Institute.
Social housing of large primates: Methodology for refinement of husbandry and management. 2004. Wolfensohn S. ATLA 32:149-151.
The lower row monkey cage: An overlooked variable in biomedical research. 2000. Reinhardt V, Reinhardt A. J Appl Anim Welf Sci 3:141-149.
The Macaques: Studies in Ecology, Behavior, and Evolution. 1980. Lindburg DG. New York: Van Nostrand.

Training nonhuman primates using positive reinforcement techniques. 2003. Prescott MJ, Buchanan-Smith HM, eds. J Appl Anim Welf Sci 6:157-161.
What factors should determine cage sizes for primates in the laboratory? 2004. Buchanan-Smith HM, Prescott MJ, Cross NJ. Anim Welf 13(Suppl):S197-S201.

Rodents and Rabbits

A Laboratory Guide to the Anatomy of the Rabbit, 2nd ed. 1966. Craigie EH. Toronto: University of Toronto Press.
Anatomy and Embryology of the Laboratory Rat. 1986. Hebel R, Stromberg MW. Wörthsee: BioMed Verlag.
Anatomy of the Guinea Pig. 1975. Cooper G, Schiller AL. Cambridge MA: Harvard University Press.
Anatomy of the Rat. 1970. Greene EC. New York: Hafner.
Bensley's Practical Anatomy of the Rabbit, 8th ed. 1948. Craigie EH, ed. Philadelphia: Blakiston.
Biology of the House Mouse. 1981. Symposia of the Zoological Society of London, No. 47. Berry RJ, ed. London: Academic Press.
Chronic stress coping in isolated and socially housed male and female rats. 2003. Westenbroek C, Den B, Gerrits M, Ter H. Society for Behavioral Neuroendocrinology Annual Meeting, Cincinnati, OH.
Comfortable quarters for gerbils in research institutions. 2002. Waiblinger E. In: Comfortable Quarters for Laboratory Animals, 9th ed. Washington: Animal Welfare Institute. p 18-25.
Comfortable quarters for guinea pigs in research institutions. 2002. Reinhardt V. In: Comfortable Quarters for Laboratory Animals, 9th ed. Washington: Animal Welfare Institute. p 38-42.
Comfortable quarters for hamsters in research institutions. 2002. Kuhnen G. In: Comfortable Quarters for Laboratory Animals, 9th ed. Washington: Animal Welfare Institute. p 33-37.
Comfortable quarters for mice in research institutions. 2002. Sherwin CM. In: Comfortable Quarters for Laboratory Animals, 9th ed. Washington: Animal Welfare Institute. p 6-17.
Comfortable quarters for rabbits in research institutions. 2002. Boers K, Gray G, Love J, Mahmutovic Z, McCormick S, Turcotte N, Zhang Y. In: Comfortable Quarters for Laboratory Animals, 9th ed. Washington: Animal Welfare Institute. p 43-49.
Comfortable quarters for rats in research institutions. 2002. Lawlor M. In: Comfortable Quarters for Laboratory Animals, 9th ed. Washington: Animal Welfare Institute. p 26-32.
Definition, Nomenclature, and Conservation of Rat Strains. 1992. Committee on Rat Nomenclature. ILAR News 34(4):S1-S24.
Effect of ambient temperature on cardiovascular parameters in rats and mice: A comparative approach. 2004. Swoap SJ, Overton JM, Garber G. Am J Physiol 287:R391-R396.
Effect of cage bedding on temperature regulation and metabolism of group-housed female mice. 2004. Gordon CJ. Comp Med 54:63-68.
Effect of temperature on the behavioural activities of male mice. 2003. Ajarem J, Ahmad M. Dirasat Pure Sci 30:59-65.
Effects of caging type and animal source on the development of foot lesions in Sprague Dawley rats (*Rattus norvegicus*). 2001. Peace TA, Singer, Niemuth NA, Shaw ME. Contemp Top Lab Anim Sci 40:17-21.
Estimates of appropriate number of rats: Interaction with housing environment. 2001. Mering S, Kaliste-Korhonen E, Nevalainen T. Lab Anim 35:80-90.

From house mouse to mouse house: The behavioural biology of free-living *Mus musculus* and its implications in the laboratory. 2004. Latham M, Mason G. Appl Anim Behav Sci 86:261-289.

Group housing and enrichment cages for breeding, fattening and laboratory rabbits. 1992. Stauffacher M. Anim Welf 1:105.

Group housing for male New Zealand White rabbits. 1997. Raje S, Stewart KL. Lab Anim 26:36-37.

Guidelines for the Well-Being of Rodents in Research. 1990. Guttman HN, ed. Bethesda MD: Scientists Center for Animal Welfare.

Guinea Pigs: Good Practice for Housing and Care, 2nd ed. 2008. RSPCA, Research Animals Department. Available at http://content.www.rspca.org.uk/cmsprd/Satellite?blobcol=urldata&blobheader=application%2Fpdf&blobkey=id&blobnocache=false&blobtable=MungoBlobs&blobwhere=1232988745157&ssbinary=true; accessed January 24, 2010.

Hamsters: Good Practice for Housing and Care, 2nd ed. 2008. RSPCA, Research Animals Department. Available at http://content.www.rspca.org.uk/cmsprd/Satellite?blobcol=urldata&blobheader=application%2Fpdf&blobkey=id&blobnocache=false&blobtable=MungoBlobs&blobwhere=1232988745472&ssbinary=true; accessed January 24, 2010.

Handbook on the Laboratory Mouse. 1975. Crispens CG Jr. Springfield IL: Charles C Thomas.

Histological Atlas of the Laboratory Mouse. 1982. Gude WD, Cosgrove GE, Hirsch GP. New York: Plenum.

Individually ventilated cages: Beneficial for mice and men? 2002. Baumans V, Schlingmann F, Vonck M, van Lith HA. Contemp Top Lab Anim Sci 41:13-19.

Individually ventilated microisolation cages. 1997. Novak G. Lab Anim 26:54-57.

Inventory of the behaviour of New Zealand white rabbits in laboratory cages. 1995. Gunn D. Appl Anim Behav Sci 45 (3/4):277-292.

Joint Working Group on Refinement: Husbandry refinements for rats, mice, dogs and non-human primates used in telemetry procedures. 2004. Seventh report of the BVAAWF/FRAME/RSPCA/UFAW Joint Working Group on Refinement, Part B. Hawkins P, Morton DB, Beyan R, Heath K, Kirkwood J, Pearce P, Scott L, Whelan G, Webb A. Lab Anim 38:1-10. Available at www.rspca.org.uk/ImageLocator/LocateAsset?asset=document&assetId=1232712323251&mode=prd; accessed August 10, 2010.

Laboratory Anatomy of the Rabbit, 3rd ed. 1990. McLaughlin CA, Chiasson RB. New York: McGraw-Hill.

Laboratory Animal Management: Rodents. 1996. National Research Council. Washington: National Academy Press.

Laboratory Hamsters. 1987. van Hoosier GL, McPherson CW, eds. New York: Academic Press.

Modulation of aggression in male mice: Influence of cage cleaning regime and scent marks. 2000. van Loo PLP, Kruitwagen CLJJ, van Zutphen LFM, Koolhaas JM, Baumans V. Anim Welf 9:281-295.

Observations on the prevalence of nest-building in non-breeding TO strain mice and their use of two nesting materials. 1997. Sherwin CM. Lab Anim 31:125-132.

Origins of Inbred Mice. 1979. Morse HC III, ed. New York: Academic Press.

Period length of the light-dark cycle influences the growth rate and food intake in mice. 1999. Campuzano A, Cambras T, Vilaplana J, Canal M, Carulla M, Diez-Noguera A. Physiol Behav 67:791-797.

Preference of guinea pigs for bedding materials: Wood shavings versus paper cutting sheet. 2003. Kawakami K, Takeuchi T, Yamaguchi S, Ago A, Nomura M, Gonda T, Komemushi S. Exper Anim (Japanese Assoc Lab Anim Sci) 52:11-15.

Preferences for nesting material as environmental enrichment for laboratory mice. 1997. VandeWeerd HA, van Loo PL, van Zutphen LF, Koolhaas JM, Baumans V. Lab Anim 31:133-143.

Preferences of laboratory mice for characteristics of soiling sites. 1996. Sherwin CM. Anim Welf 5:283-288.

Proceedings of the Third International Workshop on Nude Mice. 1982. Reed ND, ed. vol 1: Invited Lectures/Infection/Immunology; vol 2: Oncology. New York: Gustav Fischer.

Rabbits: Good Practice for Housing and Care, 2nd ed. 2008. RSPCA, Research Animals Department. Available at http://content.www.rspca.org.uk/cmsprd/Satellite?blobcol=urldata&blobheader=application%2Fpdf&blobkey=id&blobnocache=false&blobtable=MungoBlobs&blobwhere=1232988745180&ssbinary=true; accessed January 24, 2010.

Rats: Good Practice for Housing and Care, 2nd ed. 2008. RSPCA, Research Animals Department. Available at http://content.www.rspca.org.uk/cmsprd/Satellite?blobcol=urldata&blobheader=application%2Fpdf&blobkey=id&blobnocache=false&blobtable=MungoBlobs&blobwhere=1232988745061&ssbinary=true; accessed January 24, 2010.

Refinements in rabbit husbandry. 1993. Morton DB, Jennings M, Batchelor GR, Bell D, Birke L, Davies K, Eveleigh JR, Gunn D, Heath M, Howard B, Koder P, Phillips J, Poole T, Sainsbury AW, Sales GD, Smith DJA, Stauffacher M, Turner RJ. Lab Anim 27:301-329.

Refining Rabbit Care: A Resource for Those Working with Rabbits in Research. 2008. Hawkins P, Hubrecht R, Bucknell A, Cubitt S, Howard B, Jackson A, Poirier GM. RSPCA. Available at www.rspca.org.uk/ImageLocator/LocateAsset?asset=document&assetId=1232712644330&mode=prd; accessed January 24, 2010.

Refining rodent husbandry: The mouse. 1998. Jennings M, Batchelor GR, Brain PF, Dick A, Elliott H, Francis RJ, Hubrecht RC, Hurst JL, Morton DB, Peters AG, Raymond R, Sales GD, Sherwin CM, West C. Lab Anim 32:233-259.

Report of the 2002 RSPCA/UFAW rodent welfare group meeting: Individually ventilated cages and rodent welfare. 2003. Hawkins P, Anderson D, Applebee K, Key D, Wallace J, Milite G, MacArthur-Clark J, Hubrecht R, Jennings M. Anim Technol Welf 2:23-24.

Research Techniques in the Rat. 1982. Petty C. Springfield IL: Charles C Thomas.

Rodents and Rabbits: Current Research Issues. 1994. Niemi SM, Venable JS, Guttman JN, eds. Bethesda MD: Scientists Center for Animal Welfare.

Short-term effects of a disturbed light-dark cycle and environmental enrichment on aggression and stress-related parameters in male mice. 2004. Van der Meer E, van Loo PL, Baumans V. Lab Anim 38:376-383.

The Biology of the Guinea Pig. 1976. Wagner JE, Manning PJ, eds. New York: Academic Press.

The Biology of the Laboratory Rabbit, 2nd ed. 1994. Manning PJ, Ringler DH, Newcomer CE, eds. San Diego: Academic Press.

The Brattleboro rat. 1982. Sokol HW, Valtin H, eds. Ann NY Acad Sci 394:1-828.

The cage preferences of laboratory rats. 2001. Patterson-Kane EG, Harper DN, Hunt M. Lab Anim 35:74-79.

The effects of different rack systems on the breeding performance of DBA/2 mice. 2003. Tsai PP, Oppermann D, Stelzer HD, Mahler M, Hackbarth H. Lab Anim 37:44-53.

The effects of feeding and housing on the behaviour of the laboratory rabbit. 1999. Krohn TC. Lab Anim 33:101-107.

The effects of group housing on the research use of the laboratory rabbit. 1993. Whary M, Peper R, Borkowski G, Lawrence W, Ferguson F. Lab Anim 27:330.

The effects of intracage ventilation on microenvironmental conditions in filter-top cages. 1992. Lipman NS, Corning BF, Coiro MA Sr. Lab Anim 26:206-210.

The Hamster: Reproduction and Behavior. 1985. Siegel HI, ed. New York: Plenum Press.

The impact of cage ventilation on rats housed in IVC systems. 2003. Krohn TC, Hansen AK, Dragsted N. Lab Anim 37:85-93.

The impact of low levels of carbon dioxide on rats. 2003. Krohn TC, Hansen AK, Dragsted N. Lab Anim 37:94-99.

The importance of learning young: The use of nesting material in laboratory rats. 2004. van Loo PLP, Baumans V. Lab Anim 38:17-24.

The Laboratory Mouse. 2001. Danneman P, Brayton C, Suckow MA, eds. Boca Raton FL: CRC Press.

The Laboratory Mouse: Selection and Management. 1970. Simmons ML, Brick JO. Englewood Cliffs NJ: Prentice-Hall.

The laboratory rabbit. 1999. Batchelor GR. In: Poole TB, ed. UFAW Handbook on the Care and Management of Laboratory Animals, 7th ed, vol 2. Oxford: Wiley-Blackwell.

The Laboratory Rat, 2nd ed. 2005. Suckow MA, Weisbroth SH, Franklin CL. New York: Academic Press. Available at http://198.81.200.2/science/book/9780120749034; accessed January 24, 2010.

The Mouse in Biomedical Research, 2nd ed, vol I: History, Wild Mice, and Genetics. 2007. Fox JG, Barthold SW, Davisson MT, Newcomer CE, Quimby FW, Smith AL, eds. New York: Academic Press.

The Mouse in Biomedical Research, 2nd ed, vol III: Normative Biology, Husbandry, and Models. 2007. Fox JG, Barthold SW, Davisson MT, Newcomer CE, Quimby FW, Smith AL, eds. New York: Academic Press.

The Mouse in Biomedical Research, 2nd ed, vol IV: Immunology. 2007. Fox JG, Barthold SW, Davisson MT, Newcomer CE, Quimby FW, Smith AL, eds. New York: Academic Press.

The Nude Mouse in Experimental and Clinical Research, vol 1, 1978; vol 2, 1982. Fogh J, Giovanella BC, eds. New York: Academic Press.

The Rabbit: A Model for the Principles of Mammalian Physiology and Surgery. 1979. Kaplan HN, Timmons EH. New York: Academic Press.

Other Animals

Aquatic animals as models in biomedical research. 1983. Stoskopf MK. ILAR News 26:22-27.

Captive Invertebrates: A Guide to Their Biology and Husbandry. 1992. Frye FL. Malabar FL: Krieger Publishing.

Effects of experience and cage enrichment on predatory skills of black-footed ferrets (*Mustela nigripes*). 1999. Vargas A, Anderson SH. J Mammal 80:263-269.

Growth, feed efficiency and condition of common octopus (*Octopus vulgaris*) fed on two formulated moist diets. 2008. Cerezo Valverde J, Hernández MD, Aguado-Giménez F, García García B. Aquaculture 275:266-273.

Handbook of Marine Mammals. 1991. Ridgway SH, Harrison RJ, eds. New York: Academic Press.

Influence of diet on growing and nutrient utilization in the common octopus (*Octopus vulgaris*). 2002. García García B, Aguado-Giménez F. Aquaculture 211:171-182.

Laboratory Animal Management: Marine Invertebrates. 1981. National Research Council. Washington: National Academy Press.

Nontraditional Animal Models for Laboratory Research. 2004. ILAR J 45(1).

- Octodon degus: A diurnal, social, and long-lived rodent. 2004. Lee TM. ILAR J 45:14-24.
- The musk shrew (*Suncus murinus*): A model species for studies of nutritional regulation of reproduction. 2004. Temple JL. ILAR J 45:25-34.

The prairie vole: An animal model for behavioral neuroendocrine research on pair bonding. 2004. Aragona BJ, Wang ZX. ILAR J 45:35-45.
Rearing of *Octopus vulgaris* paralarvae: Present status, bottlenecks and trends. 2007. Iglesias J, Sánchez FJ, Bersano JGF, Carrasco JF, Dhont J, Fuentes L, Linares F, Muñoz JL, Okumura S, Roo J, van der Meeren T, Vidal EAG, Villanueva R. Aquaculture 266:1-15.
The Care and Management of Cephalopods in the Laboratory. 1991. Boyle PR. Herts UK: Universities Federation for Animal Welfare.
The laboratory opossum (*Monodelphis domestica*) in laboratory research. 1997. VandeBerg JL, Robinson ES. ILAR J 38:4-12.
The Marine Aquarium Reference: Systems and Invertebrates. 1989. Moe MA. Plantation FL: Green Turtle Publications.
The Principal Diseases of Lower Vertebrates. 1965. Reichenbach Klinke H, Elkan E. New York: Academic Press.
Xiphophorus interspecies hybrids as genetic models of induced neoplasia. Walter RB, Kazianis S. 2001. ILAR J 42:299-304, 309-321.

VETERINARY CARE

Transportation

Acclimatization of rats after ground transportation to a new animal facility. 2007. Capdevila S, Giral M, Ruiz de la Torre JL, Russell RJ, Kramer K. Lab Anim 41:255-261.
Effects of air transportation cause physiological and biochemical changes indicative of stress leading to regulation of chaperone expression levels and corticosterone concentration. 2009. Shim S, Lee S, Kim C, Kim B, Jee S, Lee S, Sin J, Bae C, Woo JM, Cho J, Lee E, Choi H, Kim H, Lee J, Jung Y, Cho B, Chae K, Hwang D. Exp Anim 58:11-17.
Establishing an appropriate period of acclimatization following transportation of laboratory animals. 2006. Obernier JA, Baldwin RL. ILAR J 47:364-369.
Guidance on the transport of laboratory animals. 2005. Swallow J, Anderson D, Buckwell AC, Harris T, Hawkins P, Kirkwood J, Lomas M, Meacham S, Peters A, Prescott M, Owen S, Quest R, Sutcliffe R. Thompson K. Report of the Transport Working Group, established by the Laboratory Animal Science Association (LASA). Lab Anim 39:1-39.
Reduced behavioral response to gonadal hormones in mice shipped during the peripubertal/adolescent period. 2009. Laroche J, Gasbarro L, Herman JP, Blaustein JD. Endocrinology 150:2351-2358.
The use of radiotelemetry to assess the time needed to acclimatize guineapigs following several hours of ground transport. 2009. Stemkens-Sevens S, van Berkel K, de Greeuw I, Snoeijer B, Kramer K. Lab Anim 43:78-84.
Transportation of laboratory animals. 2010. White WJ, Chou ST, Kole CB, Sutcliffe R. UFAW Handbook on the Care and Management of Laboratory and Other Research Animals, 8th ed. Hertfordshire: Universities Federation for Animal Welfare. p 169-182.

Anesthesia, Pain, and Surgery

Anaesthesia in ferrets, rabbits, and guinea pigs. 1998. Alderton B. In: Bryden D, ed. Internal Medicine: Small Companion Animals. The TG Hungerford course for veterinarians, Proceedings 306, Stephen Roberts Lecture Theatre, University of Sydney, Australia, June 15-19. University of Sydney Post Graduate Foundation in Veterinary Science. p 241-268.

Anesthesia and analgesia. 1997. Schaeffer D. In: Kohn DF, ed. Nontraditional Laboratory Animal Species in Anesthesia and Analgesia in Laboratory Animals. San Diego: Academic Press.

Anesthesia and Analgesia in Laboratory Animals. 1997. Kohn DF, Wixson SK, White WJ, Benson GJ, eds. San Diego: Academic Press.

Anesthesia and Analgesia in Laboratory Animals, 2nd ed. 2008. Fish R, Danneman PJ, Brown M, Karas A, eds. 2008. San Diego: Academic Press.

Anesthesia for Veterinary Technicians. 2010. Bryant S, ed. Somerset NJ: Wiley-Blackwell.

Anaesthetic and Sedative Techniques for Aquatic Animals, 3rd ed. 2008. Ross L, Ross B. Somerset NJ: Wiley-Blackwell.

Animal Physiologic Surgery, 2nd ed. 1982. Lang CM, ed. New York: Springer-Verlag.

AVMA Guidelines on Euthanasia. 2007. Schaumburg IL: American Veterinary Medical Association.

Challenges of pain assessment in domestic animals. 2002. Anil SS, Anil L, Deen J. JAVMA 220:313-319.

Definition of Pain and Distress and Reporting Requirements for Laboratory Animals. 2000. National Research Council. Washington: National Academy Press.

Handbook of Veterinary Anesthesia, 4th ed. 2007. Muir WW III, Hubbell JAE. Maryland Heights MO: Mosby.

Handbook of Veterinary Pain Management. 2002. Gaynor JS, Muir W. St. Louis: Mosby.

Laboratory animal analgesia, anesthesia, and euthanasia. 2003. Hedenqvist P, Hellebrekers LJ. In: Hau J, Van Hoosier GL, eds. Handbook of Laboratory Animal Science: Essential Principles and Practices, 2nd ed. Boca Raton FL: CRC Press.

Laboratory Animal Anesthesia, 3rd ed. 2009. Flecknell PA. London: Academic Press.

Lumb and Jones' Veterinary Anesthesia and Analgesia, 4th ed. 2007. Tranquilli WJ, Thurman JC, Grimm KA, eds. San Francisco: Wiley-Blackwell.

Pain alleviation in laboratory animals: Methods commonly used for perioperative pain relief. 2002. Vainio O, Hellsten C, Voipio HM. Scand J Lab Anim Sci 29:1-21.

Pain and distress. 2006. Karas A, Silverman J. In: Suckow M, Silverman J, Murthy S, eds. The IACUC Handbook. Boca Raton FL: CRC Press.

Pain Management in Animals. 2000. Flecknell PA, Waterman-Pearson A, eds. Philadelphia: WB Saunders.

Paralytic agents. 1997. Hildebrand SV. In: Kohn DF, Wixson SK, White WJ, Benson GJ, eds. Anesthesia and Analgesia in Laboratory Animals. San Diego: Academic Press.

Position Statement on Recognition and Alleviation of Pain and Distress in Laboratory Animals. 2000. AALAS. Available at www.aalas.org/pdf/Recognition_and_Alleviation_of_Pain_and_Distress_in_Laboratory_Animals.pdf; accessed January 24, 2010.

Recognition and Alleviation of Pain in Laboratory Animals. 2009. National Research Council. Washington: National Academies Press.

Recognizing pain and distress in laboratory animals. 2000. Carstens E, Moberg GP. ILAR J 41:62-71.

Recommendations for euthanasia of experimental animals. 1996. Close B, Baniste K, Baumans V, Bernoth EM, Bromage N, Bunyan J, Erhardt W, Flecknell P, Gregory N, Hackbarth H, Morton D, Warwick C. Lab Anim 30:293-316.

Small Animal Anesthesia and Analgesia. 2008. Carroll GL. Ames IA: Blackwell Publishing.

Small Animal Surgery Textbook, 3rd ed. 2007. Fossum T. Maryland Heights MO: Mosby.

Small Animal Surgical Nursing, 2nd ed. 1994. Mosby's Fundamentals of Animal Health Technology. Tracy DL, ed. St. Louis: Mosby.

Surgery: Basic principles and procedures. 2003. Waynforth HB, Swindle MM, Elliott H, Smith AC. In: Hau J, Van Hoosier GL, eds. Handbook of Laboratory Animal Science: Essential Principles and Practices, 2nd ed, vol 1. Boca Raton FL: CRC Press. p 487-520.

Textbook of Small Animal Surgery, 3rd ed. 2003. Slatter D. Philadelphia: WB Saunders.
The Biology of Animal Stress: Basic Principles and Implications for Animal Welfare. 2000. Moberg GP, Mench JA. Wallingford UK: CAB International.
The IACUC Handbook. 2006. Flecknell P, Silverman J, Suckow MA, Murthy S, eds. New York: CRC Press.
The importance of awareness for understanding fetal pain. 2005. Mellor DJ, Diesch TJ, Gunn AJ, Bennet L. Brain Res Rev 49:455-471.
When does stress become distress? 1999. Moberg GP. Lab Anim 28:422-426.

Disease Surveillance, Diagnosis, and Treatment

Clinical Laboratory Animal Medicine. 2008. Hrapkiewicz K, Medina L, eds. San Francisco: Wiley-Blackwell.
Current strategies for controlling/eliminating opportunistic microorganisms. 1998. White WJ, Anderson LC, Geistfeld J, Martin D. ILAR J 39:391-305.
Drug Dosage in Laboratory Animals: A Handbook. 1989. Borchard RE, Barnes CD, Eltherington LG. West Caldwell NJ: Telford Press.
FELASA recommendations for the health monitoring of breeding colonies and experimental units of cats, dogs and pigs. 1998. Rehbinder C, Baneux P, Forbes D, van Herck H, Nicklas W, Rugaya Z, Winkler G. Report of the Federation of European Laboratory Animal Science Associations (FELASA) Working Group on Animal Health. Lab Anim 32:1-17.
Ferrets, Rabbits and Rodents: Clinical Medicine and Surgery. 1997. Hillyer EV, Quesenberry KE. Philadelphia: WB Saunders.
Handbook of Veterinary Drugs: A Compendium for Research and Clinical Use. 1975. Rossoff IS. New York: Springer.
Invertebrate Medicine. 2006. Lewbart GA. Ames: Blackwell Publishing.
Kirk and Bistner's Handbook of Veterinary Procedures and Emergency Treatment, 8th ed. 2006. Ford RB, Mazzaferro E. Philadelphia: WB Saunders.
Laboratory Animal Medicine. 2002. Fox JG, Anderson LC, Loew FM, Quimby FW, eds. New York: Academic Press.
Mosby's Fundamentals of Animal Health Technology: Principles of Pharmacology. 1983. Giovanni R, Warren RG, eds. St. Louis: CV Mosby.
Veterinary Applied Pharmacology and Therapeutics, 5th ed. 1991. Brander GC, Pugh DM, Bywater RJ. London: Bailliere Tindall.
Veterinary Pharmacology and Therapeutics, 6th rev. ed. 1988. Booth NH, McDonald LE. Ames: Iowa State University Press.

Pathology, Clinical Pathology, and Parasitology

An Atlas of Laboratory Animal Haematology. 1981. Sanderson JH, Phillips CE. Oxford: Clarendon Press.
An Introduction to Comparative Pathology: A Consideration of Some Reactions of Human and Animal Tissues to Injurious Agents. 1962. Gresham GA, Jennings AR. New York: Academic Press.
Animal Clinical Chemistry: A Practical Guide for Toxicologists and Biomedical Researchers, 2nd ed. 2009. Evans GO, ed. Boca Raton FL: CRC Press.
Animal Hematotoxicology: A Practical Guide for Toxicologists and Biomedical Researchers. 2009. Evans GO. Boca Raton FL: CRC Press.
Atlas of Experimental Toxicological Pathology. 1987. Gopinath C, Prentice DE, Lewis DJ. Boston: MTP Press.

Blood: Atlas and Sourcebook of Hematology, 2nd ed. 1991. Kapff CT, Jandl JH. 1991. Boston: Little and Brown.
Clinical Biochemistry of Domestic Animals, 4th ed. 1989. Kaneko JJ, ed. New York: Academic Press.
Clinical Chemistry of Laboratory Animals. 1988. Loeb WF, Quimby FW. New York: Pergamon Press.
Color Atlas of Comparative Veterinary Hematology. 1989. Hawkey CM, Dennett TB. Ames: Iowa State University Press.
Color Atlas of Hematological Cytology, 3rd ed. 1992. Hayhoe GFJ, Flemans RJ. St. Louis: Mosby Year Book.
Comparative Neuropathology. 1962. Innes JRM, Saunders LZ, eds. New York: Academic Press.
Duncan and Prasse's Veterinary Laboratory Clinical Pathology. 2003. Latimer KS, Mahaffey EA, Prasse KW. San Francisco: Wiley-Blackwell.
Essentials of Veterinary Hematology. 1993. Jain NC. Philadelphia: Lea and Febiger.
Flynn's Parasites of Laboratory Animals, 2nd ed. 2007. Baker DG, ed. Ames: Iowa State University Press.
Handbook of Laboratory Animal Bacteriology. 2000. Hanson AK. Boca Raton FL: CRC Press.
Immunologic Defects in Laboratory Animals. 1981. Gershwin ME, Merchant B, eds. New York: Plenum.
Laboratory Profiles of Small Animal Diseases. 1981. Sodikoff C. Santa Barbara: American Veterinary Publications.
Natural Pathogens of Laboratory Animals: Their Effects on Research. 2003. Baker DG. Washington: American Society for Microbiology.
Pathology of Domestic Animals, 4th ed. 1992. Jubb KVF, Kennedy PC, Palmer N, eds. 1992. New York: Academic Press.
Pathology of Laboratory Animals. 1978. Benirschke K, Garner FM, Jones TC. 1978. New York: Springer-Verlag.
The Pathology of Laboratory Animals. 1965. Ribelin WE, McCoy JR, eds. Springfield IL: Charles C Thomas.
Veterinary Clinical Parasitology, 6th ed. 1994. Sloss MW, Kemp RL. 1994. Ames: Iowa State University Press.
Veterinary Pathology, 5th ed. 1983. Jones TC, Hunt RD. Philadelphia: Lea and Febiger.

Species-Specific References—Veterinary Care

Agricultural Animals

Basic Surgical Exercises Using Swine. 1983. Swindle MM. 1983. New York: Praeger.
Diseases of Poultry, 9th ed. 1991. Calnek BW, Barnes HJ, Beard CW, Reid WM, Yoder HW, eds. Ames: Iowa State University Press.
Diseases of Sheep. 1974. Jensen R. Philadelphia: Lea and Febiger.
Diseases of Swine, 7th ed. 1992. Leman AD, Straw BE, Mengeline WL, eds. Ames: Iowa State University Press.
FELASA recommendations for the health monitoring of breeding colonies and experimental units of cats, dogs and pigs. 1998. Rehbinder C, Baneux P, Forbes D, van Herck H, Nicklas W, Rugaya Z, Winkler G. Report of the Federation of European Laboratory Animal Science Associations (FELASA) Working Group on Animal Health. Lab Anim 32:1-17.
Swine in the Laboratory: Surgery, Anesthesia, Imaging, and Experimental Techniques, 2nd ed. Swindle MM, ed. 2007. Boca Raton FL: CRC Press.

Techniques in Large Animal Surgery, 3rd ed. 2007. Hendrickson D, ed. 2007. Somerset NJ: Wiley-Blackwell.
Textbook of Large Animal Surgery, 2nd ed. 1987. Oehme FW, Prier JE. Baltimore: Williams and Wilkins.

Amphibians, Reptiles, and Fish

An evaluation of current perspectives on consciousness and pain in fishes. 2004. Chandroo KP, Yue S, Moccia RD. Fish Fisher 5:281-295.
Anesthesia and analgesia in reptiles. Mosley CA. 2005. Semin Avian Exot Pet Medic 14:243-262.
Can fish suffer? Perspectives on sentience, pain, fear and stress. 2004. Chandroo KP, Duncan IJH, Moccia RD. App Anim Behav Sci 86:225-250.
Disease Diagnosis and Control in North American Marine Aquaculture, 2nd rev ed. 1988. Sindermann CJ, Lichtner DV. New York: Elsevier.
Diseases of Fishes. 1971. Bullock GL. Book 2B: Identification of Fish Pathogenic Bacteria. Neptune NJ: TFH Publications.
Diseases of Fishes. 1971. Bullock GL, Conroy DA, Snieszko SF. Neptune NJ: TFH Publications.
Diseases of Fishes. 1974. Anderson DP. Book 4: Fish Immunology. Neptune NJ: TFH Publications.
Diseases of Fishes. 1976. Wedemeyer GA, Meyer FP, Smith L. Book 5: Environmental Stress and Fish Diseases. Neptune NJ: TFH Publications.
Do fish have nociceptors? Evidence for the evolution of a vertebrate sensory system. 2003. Sneddon LU, Braithwaite VA, Gentle MJ. Proc R Soc Lond B 270:1115-1121.
Evaluation of rapid cooling and tricaine methanesulfonate (MS-222) as methods of euthanasia in zebrafish (*Danio rerio*). 2009. JM Wilson, RM Bunte, AJ Carty. JAALAS 48:785-789.
Evaluation of the use of anesthesia and analgesia in reptiles. 2004. Read MR. JAVMA 227:547-552.
Fish and welfare: Do fish have the capacity for pain perception and suffering? 2004. Braithwaite VA, Huntingford FA. Anim Welf 13:S87-S92.
Fish, amphibian, and reptile analgesia. Machin KL. 2001. Vet Clin North Am Exot Anim Pract 4:19-33.
Infectious Diseases and Pathology of Reptiles: Color Atlas and Text. 2007. Jacobson E. Boca Raton FL: CRC Press.
Medicine and Surgery of Tortoises and Turtles. 2004. McArthur S, Wilkinson R, Meyer J. San Francisco: Wiley-Blackwell.
Mycobacteriosis in fishes: A review. 2009. Gauthier DT, Rhodes MW. Vet J 180:33-47.
Pain and Distress in Fish. 2009. ILAR J 50(4).
    Fish sedation, analgesia, anesthesia, and euthanasia: Considerations, methods, and types of drugs. 2009. Neiffer DL, Stamper MA. ILAR J 50:343-360.
    Morphologic effects of the stress response in fish. 2009. Harper C, Wolf JC. ILAR J 50:387-396.
    Pain perception in fish: Indicators and endpoints. 2009. Sneddon LU. ILAR J 50:338-342.
Parasites, behaviour and welfare in fish. 2007. Barber I. Appl Anim Behav Sci 104:251-264.
Parasites of Freshwater Fishes: A Review of Their Treatment and Control. 1974. Hoffman GL, Meyer FP. Neptune NJ: TFH Publications.
Recommendations for control of pathogens and infectious diseases in fish research facilities. 2009. Kent ML, Feist SW, Harper C, Hoogstraten-Miller S, Mac Lawe J, Sánchez-Morgado JM, Tanguay RL, Sanders GE, Spitsbergen JM, Whipps CM. Comp Biochem Physiol Part C 149:240-248.

The ability of clove oil and MS-222 to minimize handling stressing rainbow trout (*Oncorhynchus mykiss* Walbaum). 2003. Wagner GN, Singer TD, McKinley RS. Aquacult Res 34:1139-1146.
The efficacy of clove oil as an anaesthetic for rainbow trout, *Oncorhynchus mykiss* (Walbaum). 1998. Keene JL, Noakes DLG, Moccia RD, Soto CG. Aquacult Res 29:89-101.
The neurobehavioral nature of fishes and the question of awareness and pain. Rose JD. 2002. Rev Fish Sci 10:1-38.
The Pathology of Fishes. 1975. Ribelin WE, Migaki G, eds. Madison: University of Wisconsin.

Birds

Avian Disease Manual, 6th ed. 2006. Charlton BR, ed. American Association of Avian Pathologists. Jacksonville, FL. Available at www.aaap.info.
Clinical Avian Medicine. 2006. Harrison GJ, Lightfoot TL, eds. 2006. Sphinx Publishing.
Diseases of Wild Waterfowl, 2nd ed. 1997. Wobeser GA. New York: Plenum Press.
Field Manual of Wildlife Disease: General Field Procedures and Diseases of Birds. 1999. Friend M, Franson JC, eds. US Geological Survey. Available at www.nwhc.usgs.gov/publications/field_manual/field_manual_of_wildlife_diseases.pdf; accessed January 24, 2010.
Handbook of Avian Medicine. 2009. Tully TN, Dorrestein GM, Jones AK, eds. Philadelphia: WB Saunders.
Pigeon Health and Disease. 1991. Tudor DC. Ames: Iowa State University Press.

Cats and Dogs

FELASA recommendations for the health monitoring of breeding colonies and experimental units of cats, dogs and pigs. 1998. Rehbinder C, Baneux P, Forbes D, van Herck H, Nicklas W, Rugaya Z, Winkler G. Report of the Federation of European Laboratory Animal Science Associations (FELASA) Working Group on Animal Health. Lab Anim 32:1-17.
Textbook of Veterinary Internal Medicine: Diseases of the Dog and Cat, 6th ed. 2005. Ettinger SJ, Feldman EC, eds. Philadelphia: WB Saunders.

Exotic, Wild, and Zoo Animals

Biology, Medicine, and Surgery of South American Wild Animals. 2001. Fowler ME, Cubas ZS. Ames: Iowa State University Press.
CRC Handbook of Marine Mammal Medicine: Health, Disease, and Rehabilitation. 2001. Dierauf LA, Gulland FMD. Boca Raton FL: CRC Press.
Diseases of Exotic Animals: Medical and Surgical Management. 1983. Philadelphia: WB Saunders.
Essentials of Disease in Wild Animals. 2005. Wobeser GA. Oxford: Wiley-Blackwell.
Exotic Animal Formulary, 3rd ed. 2004. Carpenter JW. Philadelphia: WB Saunders.
Infectious Diseases of Wild Mammals. 2000. Williams ES, Barker IK, eds. Oxford: Wiley-Blackwell.
Pathology of Zoo Animals. 1983. Griner LA. San Diego: Zoological Society of San Diego.
Veterinary Clinics of North America: Exotic Animal Practice Series. Elsevier.
Zoo and Wild Animal Medicine: Current Therapy 4. Fowler ME, Miller RE, eds. 1999. Philadelphia: WB Saunders.
Zoo and Wild Animal Medicine, 5th ed. 2003. Fowler E, Miller RE, eds. Philadelphia: WB Saunders.

Zoo and Wild Animal Medicine Current Therapy, 6th ed. 2007. Fowler ME, Miller RE, eds. Philadelphia: WB Saunders.
Zoo Animal and Wildlife Immobilization and Anesthesia. 2007. West G, Heard D, Caulkett N, eds. Oxford: Wiley-Blackwell.

Nonhuman Primates

Handbook of Primate Husbandry. 2005. Wolfensohn S, Honess P. Oxford: Blackwell
Nonhuman Primates in Biomedical Research: Diseases. 1998. Bennett BT, Abee CR, Henrickson R, eds. New York: Academic Press.
Nursery Rearing of Nonhuman Primates in the 21st Century. 2005. Sackett GP, Ruppenthal GC, Elias K. Chicago: University of Chicago Press.
Primate Parasite Ecology: The Dynamics and Study of Host-Parasite Relationships. 2009. Huffman MA, Chapman CA, eds. West Nyack NY: Cambridge University Press.
Simian Virology. 2009. Voevodin AF, Marx PA. Ames IA: Wiley-Blackwell.
The Emergence of Zoonotic Diseases. 2002. Burroughs T, Knobler S, Lederberg J, eds. Institute of Medicine. Washington: National Academies Press.
Zoonoses and Communicable Diseases Common to Man and Animals. 2003. Acha PN, Szyfres B. Washington: Pan American Health Organization.

Rodents and Rabbits

A Guide to Infectious Diseases of Guinea Pigs, Gerbils, Hamsters, and Rabbits. 1974. National Research Council. Washington: National Academy of Sciences.
Anesthesia and analgesia for laboratory rodents. 2008. Gaertner D, Hallman T, Hankenson F, Batchelder M. In: Fish R, Brown M, Danneman P, Karas A, eds. Anesthesia and Analgesia in Laboratory Animals. San Diego: Academic Press. p 239-298.
Aversion to gaseous euthanasia agents in rats and mice. 2002. Leach MC, Bowell VA, Allan TF, Morton DB. Comp Med 52:249-257.
Behavioural and cardiovascular responses of rats to euthanasia using carbon dioxide gas. 1997. Smith W, Harrap SB. Lab Anim 31:337-346.
Biology and Medicine of Rabbits and Rodents. 1989. Harkness JE, Wagner JE. Philadelphia: Lea and Febiger.
Common Lesions in Aged B6C3F (C57BL/6N x C3H/HeN)F and BALB/cStCrlC3H/Nctr Mice. 1981. Registry of Veterinary Pathology, Armed Forces Institute of Pathology. Washington: Armed Forces Institute of Pathology.
Common Parasites of Laboratory Rodents and Lagomorphs: Laboratory Animal Handbook. 1972. Owen D. London: Medical Research Council.
Complications of Viral and Mycoplasmal Infections in Rodents to Toxicology Research and Testing. 1986. Hamm TE, ed. Washington: Hemisphere Publishing.
Current Strategies for Controlling/Eliminating Opportunistic Microorganisms. 1998. White WJ, Anderson LC, Geistfeld J, Martin DG. ILAR J 39:291-305.
Detection and clearance of *Syphacia obvelata* in Swiss Webster and athymic nude mice. 2004. Clarke CL, Perdue KA. Contemp Top Lab Anim Sci 43:9-13.
Effective eradication of pinworms (*Syphacia muris, Syphacia obvelata* and *Aspiculuris tetraptera*) from a rodent breeding colony by oral anthelmintic therapy. 1998. Zenner L. Lab Anim 32:337-342.
Efficacy of various therapeutic regimens in eliminating *Pasteurella pneumotropica* from the mouse. 1996. Goelz MF, Thigpen JE, Mahler J, Rogers WP, Locklear J, Weigler BJ, Forsythe DB. Lab Anim Sci 46:280-284.

Eradication of infection with *Helicobacter* spp. by use of neonatal transfer. 2000. Truett GE, Walker JA, Baker DG. Comp Med 50:444-451.

Eradication of murine norovirus from a mouse barrier facility. 2008. Kastenmayer RJ, Perdue KA, Elkins WR. JAALAS 47:26-30.

Euthanasia of neonatal mice with carbon dioxide. 2005. Pritchett K, Corrow D, Stockwell J, Smith A. Comp Med 55:275-281.

Euthanasia of rats with carbon dioxide: Animal welfare aspects. 2000. Hackbarth H, Kuppers N, Bohnet W. Lab Anim 34:91-96.

Experimental and Surgical Technique in the Rat. 1992. Waynforth HB, Flecknell PA. New York: Academic Press.

Experimental Surgical Models in the Laboratory Rat. 2009. Rigalli A, DiLoreto V, eds. Boca Raton FL: CRC Press.

Gender influences infectivity in C57BL/6 mice exposed to mouse minute virus. 2007. Thomas ML 3rd, Morse BC, O'Malley J, Davis JA, St Claire MB, Cole MN. Comp Med 57:74-81.

Guidelines for the Euthanasia of Mouse and Rat Feti and Neonates. 2007. Office of Animal Care and Use, NIH. Avaialable at: http://oacu.od.nih.gov/ARAC/documents/Rodent_Euthanasia_Pup.pdf; accessed January 24, 2010.

Helicobacter bilis-induced inflammatory bowel disease in SCID mice with defined flora. 1997. Shomer NH, Dangler CA, Schrenzel MD, Fox JG. Infect Immun 65:4858-4864.

Humane and practical implications of using carbon dioxide mixed with oxygen for anesthesia or euthanasia of rats. 1997. Danneman PJ, Stein S, Walshaw SO. Lab Anim Sci 47:376-385.

Improving murine health surveillance programs with the help of on-site enzyme-linked immunosorbent assay. 2006. Zamora BM, Schwiebert RS, Lawson GW, Sharp PE. JAALAS 45:24-28.

Infectious Diseases of Mice and Rats. 1991. National Research Council. Washington: National Academy Press.

Ivermectin eradication of pinworms from rats kept in ventilated cages. 1993. Huerkamp MJ. Lab Anim Sci 43:86-90.

Large-scale rodent production methods make vendor barrier rooms unlikely to have persistent low-prevalence parvoviral infections. 2005. Shek WR, Pritchett KR, Clifford CB, White WJ Contemp Top Lab Anim Sci 44:37-42.

Lymphocytic choriomeningitis infection undetected by dirty-bedding sentinel monitoring and revealed after embryo transfer of an inbred strain derived from wild mice. 2007. Ike F, Bourgade F, Ohsawa K, Sato H, Morikawa S, Saijo M, Kurane I, Takimoto K, Yamada YK, Jaubert J, Berard M, Nakata H, Hiraiwa N, Mekada K, Takakura A, Itoh T, Obata Y, Yoshiki A, Montagutelli X. Comp Med 57:272-281.

Microbiological assessment of laboratory rats and mice. 1998. Weisbroth SH, Peters R, Riley LK, Shek W. ILAR J 39:272-290.

Microbiological quality control for laboratory rodents and lagomorphs. 2002. Shek WR, Gaertner DJ. In: Fox, JG, Anderson LC, Loew FM, Quimby FW, eds. Laboratory Animal Medicine, 2nd ed. London: Academic Press. p 365-393.

Monitoring sentinel mice for *Helicobacter hepaticus, H. rodentium,* and *H. bilis* by use of polymerase chain reaction analysis and serological testing. 2000. Whary MT, Cline JH, King AE, Hewes KM, Chojnacky D, Salvarrey A, Fox JG. Comp Med 50:436-443.

Mouse parvovirus infection potentiates allogeneic skin graft rejection and induces syngeneic graft rejection. 1998. McKisic MD, Macy JD, Delano ML, Jacoby RO, Paturzo FX, Smith AL. Transplantation 65:1436-1446.

Pathology of Aging Rats: A Morphological and Experimental Study of the Age-Associated Lesions in Aging BN/BI, WAG/Rij, and (WAG x BN)F Rats. 1978. Burek JD. Boca Raton FL: CRC Press.

Pathology of Aging Syrian Hamsters. 1983. Schmidt RE, Eason RL, Hubbard GB, Young JT, Eisenbrandt DL. Boca Raton FL: CRC Press.

Pathology of Laboratory Rodents and Rabbits, 3rd ed. 2008. Percy DH, Barthold SW. San Francisco: Wiley-Blackwell.

Pathology of the Syrian Hamster. 1972. Homburger F. Basel NY: Karger.

Preliminary recommendations for health monitoring of mouse, rat, hamster, guinea pig, gerbil and rabbit experimental units. 1995. Hem A, Hansen AK, Rehbinder C, Voipio HM, Engh E. Scand J Lab Anim Sci 22:49-51.

Prenatal transmission and pathogenicity of endogenous ecotropic murine leukemia virus AKV. 1999. Hesse I, Luz A, Kohleisen B, Erfle V, Schmidt J. Lab Anim Sci 49:488-495.

Recommendations for the health monitoring of rodent and rabbit colonies in breeding and experimental units. 2002. Nicklas W, Baneux P, Boot R, Decelle T, Deeny AA, Fumanelli M, Illgen-Wilcke B. Lab Anim 36:20-42.

Reliability of soiled bedding transfer for detection of mouse parvovirus and mouse hepatitis virus. 2007. Smith PC, Nucifora M, Reuter JD, Compton SR. Comp Med 57:90-96.

Soiled-bedding sentinel detection of murine norovirus 4. 2008. Manuel CA, Hsu CC, Riley LK, Livingston RS. JAALAS 47:31-36.

Successful rederivation of contaminated mice using neonatal transfer with iodine immersion. 2005. Watson J, Thompson KN, Feldman SH. Comp Med 55:465-469.

Surgery of the Digestive System in the Rat. 1965. Lambert R. (Translated from the French by B. Julien). Springfield IL: Charles C Thomas.

The Mouse in Biomedical Research, 2nd ed, vol II: Diseases. 2007. Fox JG, Barthold SW, Davisson MT, Newcomer CE, Quimby FW, Smith AL, eds. New York: Academic Press.

Transfer of *Helicobacter hepaticus* infection to sentinel mice by contaminated bedding. 1998. Livingston RS, Riley LK, Besch-Williford CL, Hook RR Jr, Franklin CL. Lab Anim 48:291-293.

Viral and Mycoplasmal Infections of Laboratory Rodents: Effects on Biomedical Research. 1986. Blatt PN. Orlando: Academic Press.

DESIGN AND CONSTRUCTION OF ANIMAL FACILITIES

Air Handling Systems Ready Reference Manual. 1986. Grumman DL. New York: McGraw-Hill.

Approaches to the Design and Development of Cost-Effective Laboratory Animal Facilities. 1993. Canadian Council on Animal Care (CCAC) proceedings. Ottawa, CCAC.

Aquatic Facilities. 2008. Diggs HE, Parker JM. In: Hessler J, Lehner N, eds. Planning and Designing Animal Research Facilities. 2008. Orlando: Academic Press. p 323-331.

ARS Facilities Design Standards. 2002. USDA. Available at www.afm.ars.usda.gov/ppweb/PDF/242-01M.pdf; accessed January 24, 2010.

Biomedical and Animal Research Facilities Design Policies and Guidelines. National Institutes of Health. Available at http://orf.od.nih.gov/PoliciesAndGuidelines/BiomedicalAndAnimalResearchFacilitiesDesignPoliciesAndGuidelines/; accessed January 24, 2010.

Comfortable Quarters for Laboratory Animals, rev 1979. Washington: Animal Welfare Institute.

Control of the Animal House Environment. 1976. McSheely T, ed. 1976. London: Laboratory Animals Ltd.

Design and Management of Research Facilities for Mice. Lipman NS. 2007. In: Fox JG, Barthold SW, Davisson M, Newcomer CE, Quimby FW, Smith AL, eds. The Mouse in Biomedical Research, vol III: Normative Biology, Immunology and Husbandry. Orlando: Academic Press. p 271-319.

Design and optimization of airflow patterns. 1994. Reynolds SD, Hughes H. Lab Anim 23:46-49.

Design of Biomedical Research Facilities. 1981. Proceedings of the National Cancer Institute Symposium, National Cancer Institute. Monograph Series, vol 4. NIH Pub. No. 81-2305.

Design of surgical suites and post surgical care units. 1997. White WJ, Blum JR. In: Kohn DF, Wixson SK, White WJ, Benson GJ, eds. Anesthesia and Analgesia in Laboratory Animals. San Diego: Academic Press.

Estimating heat produced by laboratory animals. 1964. Brewer NR. Heat Piping Air Cond 36:139-141.

Guidelines for Construction and Equipment of Hospitals and Medical Facilities, 2nd ed. 1987. American Institute of Architects Committee on Architecture for Health. Washington: American Institute of Architects Press.

Guidelines for Laboratory Design: Health and Safety Considerations. 1993. DiBerardinis LJ, Baum JS, First MW, Gatwood GT, Groden EF, Seth AK. New York: John Wiley and Sons.

Handbook of Facilities Planning, vol 2: Laboratory Animal Facilities. Ruys T, ed. New York: Van Nostrand.

Laboratory Animal Houses: A Guide to the Design and Planning of Animal Facilities. 1976. Clough G, Gamble MR. LAC Manual Series No. 4. Carshalton UK: Laboratory Animals Centre.

Laboratory Animal Housing. 1978. National Research Council. Washington: National Academy of Sciences.

Livestock behavior and the design of livestock handling facilities. 1991. Grandin T. In: Ruys T, ed. Handbook of Facilities Planning, vol 2: Laboratory Animal Facilities. New York: Van Nostrand. p 96-125.

Management and Design: Breeding Facilities. 2007. White WJ. In: Fox JG, Barthold SW, Davisson MT, Newcomer CE, Quimby FW, Smith AL, eds. The Mouse in Biomedical Research, 2nd ed, vol III: Normative Biology, Husbandry, and Models. New York: Academic Press. p 235-269.

Planning and Designing Animal Research Facilities. 2008. Hessler J, Lehner N, eds. Orlando: Academic Press.

Rodent Facilities and Caging Systems. 2009. Lipman NS. In: Hessler J, Lehner N, eds. Planning and Designing Animal Research Facilities. Orlando: Academic Press. p 265-288.

Structures and Environment Handbook, 11th ed, rev 1987. Ames: Midwest Plan Service, Iowa State University.

Warning! Nearby construction can profoundly affect your experiments. 1999. Dallman MF, Akana SF, Bell ME, Bhatnagar S, Choi SJ, Chu A, Gomez F, Laugero K, Sorian L, Viau V. Endocrine 11:111-113.

Working safely at animal biosafety level 3 and 4: Facility design and management implications. 1997. Richmond JY, Ruble DL, Brown B, Jaax GP. Lab Anim 26:28-35.

APPENDIX B

U.S. Government Principles for the Utilization and Care of Vertebrate Animals Used in Testing, Research, and Training

The development of knowledge necessary for the improvement of the health and well-being of humans as well as other animals requires in vivo experimentation with a wide variety of animal species. Whenever U.S. Government agencies develop requirements for testing, research, or training procedures involving the use of vertebrate animals, the following principles shall be considered; and whenever these agencies actually perform or sponsor such procedures, the responsible Institutional Official shall ensure that these principles are adhered to:

I. The transportation, care, and use of animals should be in accordance with the Animal Welfare Act (7 U.S.C. 2131 et. seq.) and other applicable Federal laws, guidelines, and policies.[1]

II. Procedures involving animals should be designed and performed with due consideration of their relevance to human or animal health, the advancement of knowledge, or the good of society.

III. The animals selected for a procedure should be of an appropriate species and quality and the minimum number required to obtain valid results. Methods such as mathematical models, computer simulation, and in vitro biological systems should be considered.

[1]For guidance throughout these Principles, the reader is referred to the *Guide for the Care and Use of Laboratory Animals* prepared by the Institute for Laboratory Animal Research, National Academy of Sciences.

IV. Proper use of animals, including the avoidance or minimization of discomfort, distress, and pain when consistent with sound scientific practices, is imperative. Unless the contrary is established, investigators should consider that procedures that cause pain or distress in human beings may cause pain or distress in other animals.

V. Procedures with animals that may cause more than momentary or slight pain or distress should be performed with appropriate sedation, analgesia, or anesthesia. Surgical or other painful procedures should not be performed on unanesthetized animals paralyzed by chemical agents.

VI. Animals that would otherwise suffer severe or chronic pain or distress that cannot be relieved should be painlessly killed at the end of the procedure or, if appropriate, during the procedure.

VII. The living conditions of animals should be appropriate for their species and contribute to their health and comfort. Normally, the housing, feeding, and care of all animals used for biomedical purposes must be directed by a veterinarian or other scientist trained and experienced in the proper care, handling, and use of the species being maintained or studied. In any case, veterinary care shall be provided as indicated.

VIII. Investigators and other personnel shall be appropriately qualified and experienced for conducting procedures on living animals. Adequate arrangements shall be made for their in-service training, including the proper and humane care and use of laboratory animals.

IX. Where exceptions are required in relation to the provisions of these Principles, the decisions should not rest with the investigators directly concerned but should be made, with due regard to Principle II, by an appropriate review group such as an institutional animal care and use committee. Such exceptions should not be made solely for the purposes of teaching or demonstration.

APPENDIX

C

Statement of Task

The use of laboratory animals for biomedical research, testing and education is guided by the principles of the Three Rs, replacement of animals where acceptable non-animal models exist, reduction in the number of animals to the fewest needed to obtain statistically significant data, and refinement of animal care and use to minimize pain and distress and to enhance animal well-being. The *Guide for the Care and Use of Laboratory Animals* has been a critical international publication that provides information to scientists, veterinarians and animal care personnel when the decision has been made that animal use is necessary. A committee will update the 1996 version of the *Guide for the Care and Use of Laboratory Animals* (the *Guide*) to reflect new scientific information related to the issues already covered in the *Guide*, and to add discussion and guidance on new topics of laboratory animal care and use related to contemporary animal research programs.

The committee will review the scientific literature published since the release of the 1996 *Guide* and determine whether the information in the *Guide* concurs with current scientific evidence. The committee will also review the literature on new technologies related to laboratory animal care and use and determine where new guidance is necessary to ensure the best scientific outcomes and optimal animal welfare. The committee will also take into consideration all materials and discussions provided to it, including those submitted to NIH in response to the Request for Information NOT-OD-O6-011 that requested information related to the need to update the *Guide*. Where scientifically warranted, the guidance and recommendations of the 1996 *Guide* will be changed to reflect new scientific evidence, while maintaining the performance standards of the 1996 *Guide*. The committee will ensure that any recommendations in the *Guide* will be consistent with current Public Health Service Policy, the Animal Welfare Regulations, and the most recent Report of the American Veterinary Medical Association Panel on Euthanasia.

In addition to the published report, the updated *Guide* will be posted on the Internet in a pdf or equivalent format such that users will be able to search the entire document at one time.

APPENDIX

D

About the Authors

Janet C. Garber (Chair), DVM, PhD, received her Doctor of Veterinary Medicine degree from Iowa State University and her PhD in pathophysiology from the University of Wisconsin. Her experiences have included infectious disease research at the U.S. Army Medical Research Institute of Infectious Diseases (USAMRIID), primate medicine and research, GLP device and materials evaluation, and transplantation immunology. Her current interests are in the areas of laboratory animal facility management, infectious diseases, occupational health and safety, and research program management. She most recently was Vice President, Safety Assessment, at Baxter Healthcare Corporation and is now a consultant with Garber Consulting, LLC in North Carolina. Dr. Garber is currently a member of the Council on Accreditation, AAALAC, International, and previously served as Chair of the Council. She served on the ILAR Committee to Revise the *Guide for the Care and Use of Laboratory Animals* and the Committee on Occupational Health and Safety in the Care and Use of Research Animals.

R. Wayne Barbee, PhD, is Professor and Associate Director of Research at the Department of Emergency Medicine, School of Medicine, Senior VCURES (Virginia Commonwealth University Reanimation Engineering Science Center) Fellow and Chair of the IACUC at the Virginia Commonwealth University. Dr. Barbee holds a master's degree and doctorate in physiology with three decades of research involving a wide variety of animals (bats, cats, crabs, dogs, rodents, and swine) in a number of experimental settings. His research has focused on circulatory shock and resuscitation, acute and chronic rodent surgery, and analysis of rodent hemodynamics. He has been associated with IACUCs at small, medium, and large institutions for over two decades and is familiar with the oversight of animal care and use programs. He has served on multiple study sections for both the NIH and DOD. Dr. Barbee also served as an Oxford, UK 2006 fellow (recipient, VCU Harris-Manchester Award) where he examined policies, training, and security issues related to animal care and use within the UK.

Joseph T. Bielitzki, MS, DVM, is Research Manager, University of Central Florida. Dr. Bielitzki has worked with non-human primates in the laboratory environment for 20 years. Over this period he has worked with macaques (pig-tail, long-tail, Japanese, rhesus, stump-tail), baboons (yellow, green and hybrids) squirrel monkeys, capuchin monkeys, mangabeys, gibbons, chimpanzees, orangutans, bonobos, and gorillas. In the area of non-human primates his area of expertise is in enteric diseases, nursery rearing, and colony management. He has also worked with mice and rats in a variety of international facilities. He was instrumental in the writing and acceptance of the NASA Bioethical Principles for the Use of Animals in Research (NPD 8910.1). He speaks frequently on IACUC function and the importance of ethics in the use of animals. His background includes experience in academia, industry, and government in the roles of attending veterinarian, program manager, and researcher.

Leigh Ann Clayton, DVM, is Director of Animal Health at the National Aquarium in Baltimore where she also chairs the Animal Welfare Committee. Dr. Leigh Clayton has worked in the zoo/aquarium field or the exotic pet medicine field exclusively since 2000. As she has worked with animals held in aquatics systems both in recirculating fresh and salt water, she is experienced in managing disease and accomplishing preventive health programs for fishes, amphibians, and reptiles as well as birds and mammals. She is a Diplomate of the American Board of Veterinary Practitioners (Avian). Dr. Clayton has routinely used her knowledge of nitrogen cycling and the basics of a variety of life support system designs to solve health issues in these captive settings and help ensure adequate animal health. She has served on the Executive Board of the Association of Reptilian and Amphibian Veterinarians, a role that allowed her to routinely liaise with leading researchers in the amphibian medicine field.

John C. Donovan, DVM, is President of BioResources Inc. Dr. Donovan has over 30 years' experience working as a veterinarian in biomedical research and is board certified by the American College of Laboratory Animal Medicine (ACLAM). After 7 years in the U.S. Army's Medical Research and Development Command, he spent 10 years at the National Institutes of Health, becoming the Director of the National Cancer Institute's Office of Laboratory Animal Science. He began his career in the pharmaceutical industry in 1994 as Senior Director of Worldwide Laboratory Animal Resources for Rhone Poulenc-Rorer Pharmaceuticals, leading to his position as Vice President of Laboratory Animal Science and Welfare for Aventis Pharmaceuticals in 1999. In 2001 he moved to Wyeth Pharmaceuticals where he was Vice President of BioResources until his retirement in 2007. During his career, Dr. Donovan served in numerous professional leadership positions including President of ACLAM and President of the Board of Directors of the Pennsylvania Society for Biomedical Research (PSBR). He also served on several boards including those of the National Association for Biomedical Research, ACLAM, PSBR, and the New Jersey Association for Biomedical Research.

Dennis F. Kohn, DVM, PhD, is Professor Emeritus of Clinical Comparative Pathology at Columbia University. He received his DVM from Ohio State University and a doctorate in Medical Microbiology from West Virginia University. He is board certified by the American College of Laboratory Animal Medicine. He has directed laboratory animal resource/comparative medicine programs at West Virginia University Medical Center, University of Texas Medical School at Houston, and the Health Sciences Division, Columbia University. His research interests have dealt primarily with the pathogenicity of *Mycoplasma pulmonis* in the respiratory tract of laboratory rats, and the experimental pathology induced within the CNS and joints of rats inoculated with *M. pulmonis*. He is a past president of the American College of Laboratory Animal Medicine and the American Society of Laboratory Animal Practitioners, and as chair of a number of American Association of Laboratory Animal Science committees. He has served as a Council member of AAALAC-International, as a member of the 1986 American Veterinary Medical Association's Panel on Euthanasia, and the 1996 Institute of Laboratory Animal Resources' committee to revise the *Guide*.

Neil S. Lipman, VMD, is Director of the Center of Comparative Medicine and Pathology serving the Memorial Sloan-Kettering Cancer Center (MSKCC) and the Weill Medical College of Cornell University and is Professor of Veterinary Medicine in Pathology and Laboratory Medicine at Weill Cornell as well as a Laboratory Member at the Sloan-Kettering Institute at MSKCC. Dr. Lipman is a Diplomate of the American College of Laboratory Animal Medicine with over 25 years' experience in laboratory animal medicine and science. Dr. Lipman has expertise in vivarium design, engineering, and operations, having designed over 1.5 million gross square feet of vivarium space and overseen the operation of a number of major academic animal resource programs. His research interests are principally translational and include development and analysis of new technologies especially with respect to rodent housing systems and monoclonal antibody production, the characterization of various animal models, understanding the etiopathogenesis of endocrinologic disorders affecting laboratory animal species, and development and analysis of novel therapeutic strategies. Throughout his career, Dr. Lipman has been extensively involved in the postgraduate training of laboratory animal specialists.

Paul Locke, MPH, JD, DrPH, an environmental health scientist and attorney, is Associate Professor at the Johns Hopkins University Bloomberg School of Public Health in the Department of Environmental Health Sciences, Division of Toxicology. He holds an MPH from Yale University School of Medicine, a DrPH from the Johns Hopkins University Bloomberg School of Public Health and a JD degree from Vanderbilt University School of Law. Prior to joining the Department of Environmental Health, he was the Deputy Director of the Pew Environmental Health Commission and the Director of the Center for Public Health and Law at the Environmental Law Institute. Dr. Locke's research and practice focus on how decision makers use environmental health science and toxicology in regulation and policy making and how environmental health sciences influence the policy-making process. His areas of study include alternatives to animal testing in biomedical research, with particular emphasis on toxicity testing. He also maintains an active research program in radiation studies and radiation protection policy. Dr. Locke directs the Doctor of Public Health degree program in the Department of Environmental Health Sciences and is co-director of the Johns Hopkins certificate program in Humane Science and Toxicology. From 2004 until 2009 he was a member of the National Academy of Sciences Nuclear and Radiation Studies Board, and has served on five National Academy of Sciences/National Research Council expert committees. He is admitted to practice law before the bars of the states of New York and New Jersey, the District of Columbia, the United States Court of Appeals for the Second Circuit and the United States Supreme Court.

The Honorable John Melcher, DVM, a graduate of the College of Veterinary Medicine of the Iowa State University, was a practicing veterinarian in the state of Montana until 1969, in which year he was elected to the U.S. House of Representatives. He served as a Congressman for 8 years and as a Senator for 12 years. In both the House and the Senate, Senator Melcher was noted for his interest in agriculture, protection of public lands, notably Forest Service and Bureau of Land Management lands, and animal welfare and animal health protections. In 1984 he contributed to the Animal Welfare Act with an amendment requiring consideration of the psychological well being of primates used in medical research. After retiring from Congress, Senator Melcher established a second career as a consultant for the

American Veterinary Medical Association and the American Association of Veterinary Medical Colleges. Senator Melcher represents the public's perspective.

Fred W. Quimby, VMD, PhD, is a board-certified laboratory animal veterinarian, with a doctorate in pathology, specialized in the assessment of immune function in animals. Prior to his retirement in 2007 he was Associate Vice President at Rockefeller University, while over the past 35 years he oversaw the research animal programs at three Universities (Tufts, Cornell, and Rockefeller) and held the position of Professor at Cornell's Colleges of Medicine and Veterinary Medicine. He conducted research and lectured in the fields of immunology, pathology and environmental toxicology where he focused his research on toxic shock syndrome, environmental intoxication with polychlorinated biphenyls (PCBs), and immune dysfunction in pet dogs. As a laboratory animal professional he has designed and overseen the construction of five animal facilities for research animals and a zoological park. Dr. Quimby has had broad experience with a wide assortment of laboratory animals including rodents, dogs, primates, livestock, poultry, and fish, and published on the diseases, care, and /or housing of each of them. He has served on various NAS/NRC committees including the *Guide* committee, the Committee on Immunologically Compromised Rodents (Chair), the Transgenic Animal Committee, the Committee to Develop Standards for Dogs (Chair), the Monoclonal Antibody Production Committee, and the Committee Evaluating Increasing Veterinary Involvement in Biomedical Research. He served as a member of the ILAR Council and chaired the Editorial Committee of the ILAR News. He is currently serving as a member of the Committee to Assess the Current and Future Workforce Needs in Veterinary Medicine. He was a charter member of the Society for Veterinary Ethics, a member of the Board of Directors for the National Association for Biomedical Research and a member of the Strategic Planning Committee for AAALAC International.

Patricia V. Turner, MS, DVM, DVSc, is Associate Professor and Program Leader of Laboratory Animal Science in the Department of Pathobiology at the University of Guelph, Canada, where she also is currently Chair of the Animal Care Committee. She has a doctorate in comparative pathology and is a Diplomate of both the American College of Laboratory Animal Medicine and the American Board of Toxicology. Dr. Turner has experience managing Canadian Council of Animal Care–compliant animal facilities that house a full range of species (fish, rodents and rabbits, dogs and cats, swine and sheep, nonhuman primates) in both academic and industry (GLP) settings. Dr. Turner serves as an AAALAC international ad hoc specialist with excellent knowledge of current US guidelines and regulations concerning research animal care and use. Her research interests include innate immunity and infectious disease, toxicologic pathology, and the interactions between rodents and their environment as they relate to disease susceptibility. In 2007 she was the inaugural recipient of the North American Animal Welfare Award, co-sponsored by Procter & Gamble and the Humane Society of the United States.

Geoffrey A. Wood, DVM, PhD, DVSc, is Associate Professor in the Department of Pathobiology at the Ontario Veterinary College, University of Guelph, Canada. Dr. Wood has a doctorate in cancer biology and a doctorate in veterinary pathology. He has been involved in design or pathologic characterization of hundreds of genetically engineered rodents, both as the former Associate Director of Pathology at the Centre for Modeling Human Disease in Toronto, and in his current position. His lab conducts research on cancer genetics and the process of metastasis, with a focus on bone and prostate cancer. Dr. Wood's research collaborations include projects investigating various aspects of a wide range of different cancer types, as well as studies on stem cell biology, immunity, and inflammation.

Hanno Würbel, Dr.sc.nat, is Professor of Animal Welfare and Ethology at the Justus Liebig University in Giessen, Germany. He has studied biology (zoology) at the University of Berne, Switzerland and graduated from the ETH Zürich, Switzerland with a doctorate in natural sciences. He has experience in animal behavior and in the scientific assessment of animal well-being, and has mostly worked with rodents, but also with rabbits, dogs, poultry, and horses. His research focuses on environment-dependent plasticity of brain and behavior in relation to questions of animal husbandry and animal welfare. In 2005 Dr. Würbel received the Hessian Animal Welfare Research Prize and in 2009 the Felix Wankel Animal Welfare Research Award. He is a member of the Animal Welfare Council of the German government, Central Animal Welfare Officer of the University of Giessen, and head of the University's Central Animal Facility. He is also a council member of the International Society of Applied Ethology (ISAE), editor of the *Journal of Applied Animal Welfare Science,* and editorial board member of the journal *Applied Animal Behaviour Science.*

Index

B

C

D

E

F

G

H

N

O

U

V

W

X

Z